# LIFE IS SIMPLE

# LIFE IS SIMPLE

## HOW OCCAM'S RAZOR
## SET SCIENCE FREE
## AND SHAPES THE UNIVERSE

---

## JOHNJOE MCFADDEN

BASIC BOOKS

NEW YORK

*Dedicated to Pen and Ollie who helped to keep me sane*

Basic Books
Hachette Book Group
1290 Avenue of the Americas, New York, NY 10104
www.basicbooks.com

Printed in the United States of America

Originally published in 2021 in Great Britain by Basic Books London, An imprint of John Murray Press

First US Edition: September 2021

Published by Basic Books, an imprint of Perseus Books, LLC, a subsidiary of Hachette Book Group, Inc. The Basic Books name and logo is a trademark of the Hachette Book Group.

The Hachette Speakers Bureau provides a wide range of authors for speaking events. To find out more, go to www.hachettespeakersbureau.com or call (866) 376-6591.

The publisher is not responsible for websites (or their content) that are not owned by the publisher.

Typeset in Janson Text by Palimpsest Book Production Ltd, Falkirk, Stirlingshire

Library of Congress Control Number: 2021937832

ISBNs: 9781541620445 (hardcover); 9781541620438 (ebook)

LSC-C

Printing 1, 2021

# Contents

CONTENTS

## PART IV: The Cosmic Razor

# Introduction

It is May 1964. Two American physicists stand beside a scientific instrument the size of a truck. The device, shaped like a giant ear trumpet, tops a low hillside above the small township of Holmdel in New Jersey. Both men are in their mid-thirties. Arno Penzias, born to a Bavarian Jewish family who fled to the Bronx in 1939, is tall and bespectacled with receding hair. Robert Woodrow Wilson from Houston, Texas, also tall, is dark-bearded and bald. The pair met at a conference only two years earlier. Penzias endlessly talkative, Wilson shy and tentative, hit it off. They joined forces at the world-famous Bell Laboratories to work on a project to map the stars with microwaves. Both stare at the sky. Both are baffled.

FIGURE 1: Bell Telephone Laboratories' horn antenna in Holmdel, New Jersey, with Arno Penzias and Robert Wilson.

Microwaves, radiation with wavelengths anywhere between a millimetre and a metre, had been discovered nearly a century earlier and became a hot topic when Second World War military scientists attempted to harness them for radar, and tried to make rayguns capable of shooting down enemy missiles. After the war, telecommunications companies took an interest after the physicist Robert H. Dicke, working at the world-famous Massachusetts Institute of Technology (MIT), designed an efficient receiver capable of detecting microwaves. With both emitter and detector technology available, a new means of wireless communication was on the cards.

In 1959, Bell Laboratories built the Holmdel horn antenna to detect microwaves bounced off satellites. However, interest waned and shifted to alternative wireless communication technologies so Bell took to lending out the antenna to scientists who could make good use of a giant microwave trumpet. Penzias and Wilson planned to map the sky. On 20 May 1964 they climbed into the control room, a kind of elevated garden shed connected to the rear end of the trumpet, and pointed the antenna at the sky. Yet, wherever they looked, even when they aimed the giant antenna at dark regions of the night sky with very few stars, they detected only a low background noise, a static or hiss.[1] The two men were baffled.

Their first guess was that it was some kind of interference from a local source of microwaves. They checked out, and eliminated, New York City, nuclear tests, a nearby military facility and atmospheric disturbances. Crawling inside their antenna, they even discovered a pair of roosting pigeons and suspected that their droppings might be the culprit. They set traps and cleaned up their droppings but, when the birds kept returning, the scientists resorted to shooting them. Even after the avian cull, wherever they pointed their instrument in the dark night sky, it continued to hiss, uniformly, back at them.

Princeton University is about an hour's drive from Holmdel. After the war, Robert Dicke had moved there to teach and lead a research group focused on particle physics, lasers and cosmology. His lab specialised in developing sensitive instruments to test cosmological predictions of Einstein's general theory of relativity. Cosmology was then being contested by two rival groups of theorists who vied to account for Edwin Hubble's astonishing discovery, several decades earlier, that the universe is expanding. One camp favoured the steady-state theory, which claimed that the universe had always been expanding, balanced by a continuous creation of new matter into its spaces. The rival theorists, a group that included Dicke, took the expansion at its face value and ran it backwards in time to propose that, about 14 billion years ago, the universe must have burst into existence in a cataclysmic explosion from a very tiny point.

The problem was that it wasn't easy to distinguish the rival theories as both made very similar predictions. Nevertheless, Dicke realised that an exploding universe should have left a kind of smoking cosmic gun as a uniform cloud of low energy microwave radiation. He recognised that the kind of radar detectors he had developed at MIT could be adapted to detect the cosmic energy cloud. The microwave radiation would however be very faint, far dimmer than any known radio or radar signal. Its detection would require a new generation of highly sensitive microwave detectors. Dicke and his Princeton group set out to build one.

Over the months and years, members of the group presented research talks describing their steady progress. A colleague of Penzias and Wilson attended one of these meetings and passed on news about the Princeton team's efforts to the pair. Could the horn antenna's persistent microwave hiss be the signal that Dicke was looking for? Penzias decided to give Robert Dicke a call. It came through when Dicke was having a 'brown bag lunch' meeting in his office at Princeton. His colleagues remember Dicke picking up the call and listening intently, occasionally repeating phrases such as 'horn antenna' or 'excess noise' and nodding. Finally, putting down the receiver, he turned to his group and said, 'Well, boys, we've been scooped.' Dicke realised that Penzias and Wilson had discovered the Big Bang.

The next day, Dicke and his team drove to the Bell laboratories, to admire the horn antenna and take a closer look at the data. They returned convinced that Penzias and Wilson had indeed discovered the microwave remnant of the Big Bang. What most impressed both teams was the smoothness of the cosmic microwave background

(CMB), as it was later called. It had, as far as they could tell, exactly the same intensity wherever they looked in the sky. Their discovery earned Penzias and Wilson the Nobel Prize in 1978. About a decade later, NASA launched their Cosmic Background Explorer (COBE) satellite to provide more precise measurements and discovered faint ripples, with variations in radiation intensity of less than one part in 100 thousand, in the CMB. That is a lot less than the variation in whiteness you would see in the cleanest, whitest sheet of paper that you have ever seen. A decade later, in 1998, the European Space Agency (ESA) launched their own microwave detector into space, the Planck Space Observatory, and confirmed both the faint ripples and the extraordinary uniformity of the CMB.

FIGURE 2: Cosmic microwave background.

The CMB is a kind of photograph taken of the universe when it was less than the size of the Milky Way. Its uniformity tells us that, at that moment, when the first blast of light emerged from its trillions of atoms, our universe was simple. In fact, the CMB remains the simplest object that we know of today; simpler even than a single atom. It can be described by just a single number, 0.00001, which refers to the degree of variation in its ripple intensity. As Neil Turok, director emeritus of the Perimeter Institute for Theoretical Physics in Ontario, Canada, recently commented, the CMB tells us that 'the universe turns out to be stunningly simple . . . [so much so that] we don't know how nature got away with it.'[2]

The universe *remembers* its simple beginnings so that, 14 billion years after the Big Bang, its bones remain simple. This book is about uncovering those bones – the simple building blocks of our universe – with a tool known as Occam's razor, named after a Franciscan friar called William of Occam, who lived seven centuries before Penzias and Wilson.

My own interest in simplicity began at a biology research meeting held at my workplace at the University of Surrey in the UK, around the time that ESA launched their Planck mission to measure the CMB. There I listened to a talk with the provocative title of 'Occam's razor has no place in biology' delivered by my friend and colleague Hans Westerhoff. The crux of Hans's argument was that life is too complex, even 'irreducibly complex' as Hans put it, for Occam's razor to be of any use. At that time, more than two decades ago, I knew nothing about Occam and hardly more about his razor; but I did remember that I drove past a road sign to the village of Ockham on my way to work every day. The coincidence was suffi- cient to pique my interest and persuade me to trawl through the internet that evening to see if I could find any information that might save the reputation of our locally inspired razor.

My search soon revealed that the razor was indeed named after William of Occam, born in the nearby Surrey village in the late thirteenth century. After joining the Franciscans he studied theology in Oxford where he developed his preference for the simplest solutions. This idea was not entirely new but Occam's ruthless application of the principle to dismantle much of medieval phil- osophy became so notorious that, three centuries after his death, the French theologian Libert Froidmont coined the term 'Occam's razor' to refer to William's preference for shaving away excess complexity.[3]

Today, the razor is mostly known in the form 'entities should not be multiplied beyond necessity'. 'Entities' refers to the parts of an hypothesis, explanation or model of any particular system. So, if you unexpectedly detect microwaves in your horn antenna, look for familiar entities to explain the phenomenon, such as radar facilities or pigeons, before inventing new ones, like Big Bangs. As far as we know, William never expressed his preference for parsimony in the exact form above but did express the same sentiment in phrases such as 'plurality should not be posited without necessity' or 'it is futile to do with more what can be done with less'.

In the evening that followed Hans's seminar, I pulled on the threads of William's story and, the more I pulled, the more fascinating his story became. When his ideas, including his dismantling of all the established 'proofs' of God, began to leak out of Oxford, they provoked a charge of teaching heresy and a summons to Avignon to face trial before the Pope. Yet, in Avignon, he became embroiled in an even deadlier conflict between the Pope and the Franciscans, one that provoked William to accuse the Pope of heresy and led to his flight from the city chased by a posse of papal soldiers.

This was gripping stuff but already I had sufficient ammunition to defend our local hero. In my own talk the next day, I pointed out that the razor, in its most familiar formulation, insists only that 'entities should not be multiplied beyond necessity'. The 'beyond necessity' clause is generous. If all the simpler explanations for a phenomenon fail then the razor gives you full licence to invent as many preposterous notions as you need, such as the claim that the universe popped out of an infinitesimal point of nothing 14 billion years ago, to account for your data. As Sherlock Holmes put it, 'Once you eliminate the impossible, whatever remains, no matter how improbable, must be the truth.'[4] So, to Hans's objection that the razor is too blunt an instrument to handle the delicate sinews of biology, I countered that the 'beyond necessity' clause allows us to invent as many entities as we need so long as we stop there.

The debate rumbles on between us but now runs alongside my wider fascination with William, his work and the role of his razor in science. My research has led me from the cloisters of Oxford and the palaces of Avignon to the first sparks of modern science in the medieval world. From there I followed its track as it was picked up by the giants of modern science from Copernicus to

Kepler, Newton, Einstein or Darwin, who all expressed a preference for simple solutions. The journey has persuaded me that simplicity is not just a tool of science alongside experimentation, it is as central to science as numbers are to mathematics or notes to music. Indeed, in the final analysis, simplicity is, I believe, what separates science from the countless other ways of making sense of the world. In 1934, Albert Einstein insisted that 'The grand aim of all science [is] to cover the greatest number of empirical facts by logical deduction from the smallest possible number of hypotheses or axioms.'[5] Occam's razor helps us find 'the smallest number of hypotheses or axioms'.

Nor is the work of Occam's razor done. As physics inches its way towards the simplest possible theories, biologists struggle to extract simple theories from the accelerating stream of data pouring out of genomics and other 'omics' technologies. It also remains as controversial today as it was in Occam's time. Statisticians constantly debate its value and significance. Recently a group of French scientists published a paper arguing that simple models, honed by the razor, make better sense of the Covid-19 pandemic sweeping their country than the bulky cumbersome models used by most epidemiologists. At the cutting edge of science, simplicity continues to present us with the most profound, enigmatic and sometimes unsettling insights.

Perhaps most surprisingly, it is increasingly clear that the value of Occam's razor is not limited to science. William Shakespeare insisted that 'brevity is the soul of wit' and modernity has taken that principle to heart. From the minimalist music of John Cage, to the clean architectural lines of Le Corbusier, the lean prose of Samuel Beckett or the smooth lines of the iPad, modern culture is steeped in simplicity. Occam's razor finds expression in the advice of the architect Mies van der Rohe that 'Less is more'; the computer scientist Bjarne Stroustrup's instruction to 'Make simple tasks simple', or the writer and aviator Antoine de Saint-Exupéry's observation that 'It seems that perfection is reached not when there is nothing left to add but when there is nothing left to take away.' In engineering the principle is best known under the acronym KISS, or 'Keep it simple, stupid', a design principle adopted by the US Navy in the 1960s but now universally acknowledged as fundamental to sound engineering. Occam's razor underpins the modern world.

I also want to be clear about what I am not attempting in this book. It is not my aim to provide an exhaustive history of science. Instead, I hope to convince you of the unappreciated value of Occam's razor through a selective account of key ideas and innovations that exemplify its importance and illustrate its use. This inevitably means that many significant advances made by the greatest scientists have been completely omitted. To those interested readers who might wish to fill in the gaps, I refer them to just a few of many excellent books.[6]

Moreover, and perhaps even more importantly, this book isn't so much a history of science as an account and exploration of the biggest ideas, within and outside of science, that have been inspired by Occam's razor. It begins in a world where science was essentially a branch of theology. This may seem odd to us today but, throughout most periods of human history, it has been the predominant perspective. William of Occam and his razor helped to cut science free from its theological tethers, a feat that, I believe, was crucial to the subsequent course of human history. Yet even today, science remains a prisoner of its cultural context and nowhere is this more apparent than when considering its origins and development. Accordingly, *Life is Simple* also probes the wider world where Occam's razor operates.

Lastly, there is only one science, but it has many branches and spreading roots that stretched to ancient Mesopotamia, where the first astronomers plotted the movement of the stars, and to ancient India where the system we now call Arabic numerals was invented. Those roots also reached to ancient China where many technologies such as block printing originated, to Aegean shores where ancient Greeks first used mathematics to make sense of the world, and back into the Middle East and North Africa where Islamic scholars both preserved and extended Greek science into new areas such as optics and chemistry. Hundreds of places, countless times and millions of people have contributed to that remarkable system of thought that we now call modern science. Sadly, most of the scientists whose work I have used to illustrate the role of Occam's razor are wealthy white Western men. There is no doubt that people of all genders and races have contributed to modern science, but lack of opportunity, prejudice and social barriers have largely left their contributions undocumented. I have tried to redress this deficit in the later chapters of this book to illustrate my own conviction that science has been, and will continue to be, humankind's most cooperative endeavour.

Our journey begins with a voyage.

# PART I

Discovery

# I

# Of Scholars and Heretics

I found a great many things that were heretical, erroneous, silly, ridiculous, fantastic, insane and defamatory, contrary and likewise plainly adverse to orthodox faith, good morals, natural reason, certain experience, and fraternal charity. I have decided that some of them should be inserted here.

William of Occam, 'A Letter to the Friars Minor', 1334[1]

## *Escape*

On the night of 26 May 1328, three friars, tonsured and dressed in the grey robes of the Franciscans, slipped out of the papal city of Avignon and rode south to the Crusader riverside port of Aigues-Mortes, about sixty miles north-west of Marseilles. The first was Michael of Cesena, minister general of the Franciscan order and keeper of its seal of office. The second was the Franciscans' chief lawyer, Bonagratia of Bergamo. Both were well known to princes and popes, having travelled widely between European courts as representatives of their order. The third fugitive, who was around forty and slight of build, was the English scholar William of Occam. Although more than a decade younger than his Franciscan brothers, William's dangerous ideas had already brought him notoriety and a charge of heresy. The three were fleeing papal justice after having accused the Pope of being a heretic. If captured, they would face excommunication, imprisonment or even a slow and cruel death on a burning pyre.

The group travelled with a guard of 'well-armed servants'.* At Aigues-Mortes, they were met by 'Giovanni Gentile, citizen of Savona, the captain of a galley'[2] moored in the harbour. Such ships, long and low in the water and similar in construction to a Venetian gondola but bigger and equipped with both sails and rows of oars, were able to navigate shallow seas and rivers and so were widely used to trade goods between the northern Mediterranean ports. The friars would surely have been relieved to board the galley and must have been keen to set off, but bad weather and contrary tides foiled their escape.

Meanwhile, back in Avignon, their flight had been discovered and a posse of papal soldiers had been dispatched to capture them. Led by the Lord of Arrabley and 'accompanied by a large number of papal and royal retainers', the arrest party arrived in the dead of night with the Franciscans aboard Gentile's galley still moored in the harbour unable to launch. Arrabley demanded that the ship's captain hand over the fugitives. Gentile initially appeared cooperative, inviting Lord Arrabley on board. The papal envoy formally arrested the Franciscans and threatened 'the gravest penalties' if Gentile refused to hand them over. The two men agreed a deal in which the Franciscans would be surrendered to the papal authorities. Yet after Arrabley had disembarked, and still under the cover of night, 'the captain unfurled his sails and secretly sailed away'.

Watching the angry papal soldiers recede into darkness must have delighted the terrified Franciscans. However, their glee was short-lived because, after they had 'navigated for a good thirty leagues downriver' (at this time the port was many miles from the sea), 'Divine providence created a contrary wind' that blew them back upstream, compelling Gentile to seek refuge once again within reach of the papal posse. Negotiations resumed for delivery of the Franciscans, who remained on board for several days 'in extreme

---

* This account was discovered a few decades ago in the Vatican archive by George Knysh who kindly provided a 'preliminary translation' of the Latin text. Direct quotations from the translation are indicated with quotation marks.

fear'. However, it seems that the wily captain was playing for time because, when the weather turned, he launched his boat into the river once again, this time reaching the open sea where it was met by 'a large Savonan war galley captained by a "Li Pelez"', allied to the newly elected Holy Roman Emperor Louis of Bavaria. Gentile arranged for the fugitives to transfer to the bigger ship and on Friday 3 June, the war galley and its Franciscan passengers sailed beyond the reach of the furious Pope. William lived to see another day but, as far as we know, he never returned to France or home to England.

The historical account of the Franciscans' escape breaks off after their flight from Aigues-Mortes. Yet a flavour of the kind of voyage that William and his friends would have experienced can be obtained from the near contemporary account of a departure from the same port by Jean de Joinville, who accompanied Louis IX on the seventh crusade in 1248.

> When the horses were on ship, our master mariner called to his seamen who stood on the prow and said, 'Are you ready?' and they answered, 'Aye Sir'. 'Then let the clerks and priests come forward.' As soon as they had come forward he called to them, 'Sing for God's sake!' and they all with one voice chanted, 'Veni Creator Spiritus'. Then he cried to his seamen, 'Unfurl the sails for God's sake!' and they did so. In a short space the wind had borne us out of sight of land, so that we saw nought but sky and water . . . And these things I tell you, that you may understand how foolhardy is the man who dares . . . to place himself in mortal peril, seeing that you lie down to sleep at night on shipboard, you lie down not knowing whether in the morning you may find yourself at the bottom of the sea.[3]

So, why were William's ideas so dangerous that the Pope went to so much trouble to try to catch him? To understand, we need to enter the archaic mindset of the medieval world.

◆

William was born around 1288 in Ockham, a small Surrey village about a day's ride south-west of London. There are no contemporary accounts except the village's entry in the Domesday Book written in 1086, twenty years after England's conquest by the Normans and two hundred years before William was born. This may seem a long time but, after the immediate turmoil of the Conquest, the pace of change in medieval England was much slower than today and, as far as we can tell, Ockham remained the same insignificant hamlet or village as the settlement described under its Anglo-Saxon name as Bocheham. It provided pasture for 26 cows, woodland yielding acorns to feed about 40 pigs, fields to support about 20 families and a mill. Probably the most archaic feature of the Domesday account is how the book describes the village's human inhabitants as 'thirty two villeins and four bordars . . . three bondmen'. These are all categories of serfs, who were little more than slaves required to work for their lord for no pay and were bought and sold along with the manor. None are named but one freeman with the Anglo-Saxon name of Gundrid is mentioned. The whole manor was valued at 15 pounds, which is roughly equivalent to eight times what a labourer could earn in a year.

The first concrete fact we know about William is that he was given to the Franciscan order, probably aged around eleven. This was relatively common in noble families, but several facts argue against William being of noble birth. First, there is the absence of any record of his family, suggesting that they were humble. Secondly, there are no nobles listed in Domesday's Ockham nor in later accounts. Monasteries also acted as unofficial orphanages for unwanted children left on their steps, so a more likely beginning to William's life is as an orphaned, illegitimate or abandoned child.

There were several small Franciscan friaries in towns near Ockham, for example, in Guildford and Chertsey. It is likely that William spent his early years in one of them. After arriving as a young boy, he would have been tonsured and clad in the grey hooded habit of the Franciscans.* As an oblate, a kind of apprentice friar, he would have

---

* A friar is a member of one of the mendicant orders mostly founded in the twelfth or thirteenth century and included the Carmelites, Franciscans, Dominicans

been subjected to the highly regimented order of friary life. Each day would begin at about 6 a.m. with lauds, then there would be services and singing of psalms, followed by classes. The aim of his primary education was to ensure that he grew up able to fulfil his primary duty as a friar to read prayers and sing psalms. The standard teaching method was rote-learning and chanting or singing of passages. At this stage of their education, the boys would not necessarily be expected to understand the Latin of their songs and prayers. As the boy in Chaucer's 'The Prioress's Tale' confesses, 'I learn song; I know little grammar.'

During his early years at the friary, William would have been instructed in basic arithmetic, reading the Bible and the lives of the saints. Books were very precious so teaching mostly involved rote dictation of passages read by the master and then copied onto waxed tablets with a stylus. Discipline was strictly enforced in a regime probably not too dissimilar from that advocated by St Benignus of Dijon who decreed that 'if the boys commit any fault . . . let there be no sort of delay, but let them be stripped forthwith of frock and cowl and be beaten in their shirt only.'[4] William not only survived but sufficiently impressed his superiors, so that around 1305, when he was about twenty, they sent him to the closest Franciscan school, or *studium generale*, Greyfriars near Newgate in the City of London, to receive his secondary education.

Newgate was an area in the south-east corner of the old City of London adjacent to one of the seven gates in the city walls. It lies about a day's ride north of Ockham or Guildford; or, more likely, several days' hike. The friary, the oldest and largest in England, housed over a hundred friars and was close to Newgate's busy meat market. We can imagine the novice friar elbowing his way through the noisy, slippery, stinking and bustling narrow alleys and lanes with names such as Bladder Street or the Shambles (a contraction of the 'Flesh

---

and Augustinians. Modern Franciscans wear brown but the appellation 'grey friars' is thought to derive from the early Franciscan habit (which William and his colleagues would probably have followed) of wearing robes spun from un-dyed wool that became grey with wearing. Friars were distinguished from monks, at least initially, by their adoption of a travelling hermit lifestyle but by the fourteenth century they mostly lived in friaries.

Ambles'), dodging men and boys carrying blood-dripping carcases of cows, pigs, sheep and steaming buckets of congealed blood to make the blood pudding sold on nearby Pudding Lane. It was probably with a great deal of relief that he passed through the wooden doors to arrive at the relative seclusion and quiet of the friary.

As a studium generale, Greyfriars was something between a school and university where an academically inclined friar would study for three years for a bachelor's degree or six years for a master's degree before, if he were clever enough, going on to study for a doctorate in theology. It was here that William's education would have broadened to encompass the medieval university liberal arts *trivium* comprising grammar, logic and rhetoric before advancing to the *quadrivium*, which included music as well as subjects that would today be part of a science curriculum – arithmetic, geometry and astronomy.

However, when William sat down in the stone-walled lecture room alongside his grey-robed and tonsured fellow students to listen to one of his masters lecture on logic, arithmetic, geometry or astronomy, the experience would have been wholly different from that of any modern student. For a start, most of the key texts were hundreds, even thousands, of years old.

### *The crowded cosmos before the razor*

> It seemed a cloud enclosed us, shining, dense, with polished surface firm that, diamond-bright, was dazzling in the sun's reflected light. We passed within the eternal pearl, as sinks a ray of sunlight in the stream, which drinks the light . . . If I were body or unsubstanced soul I know not . . . Faint as a white pearl on as white a brow, so there were many faces round me now eager for speech.
>
> Dante, *The Divine Comedy*, 'The sphere of the moon'

I should first point out that *science*, as the term is understood today, did not really exist in the medieval world. The word derives from the Latin *scientia* meaning knowledge. However, medieval scholars identified scientia with knowledge that could be known with certainty, such as the roundness of the moon or that the square on the hypotenuse is equal to the sum of the squares on the other two sides of a right-angled triangle. This contrasted with matters of opinion, such as whether Dante or Chaucer were the greater poet or whether theft or adultery the greater sin. Yet, in contrast to today's science, scientia also included 'theological truths' that were considered to be certain, such as the existence of Heaven and Hell.

With this clarification in mind, the first scientific (in the modern sense) scientia that William studied at Greyfriars would have been various commentaries by Greek scholars such as Euclid (mathematics) and Aristotle (mostly everything else) from the third and fourth centuries BCE and Roman scholars, such as Boethius, from the fifth–sixth century CE. By this time, Aristotle was the principal authority and William would probably have studied his *Physics*, *De Animalibus* (*On Animals*), *De Caelo et Mundo* (*Heaven and the World*), *De Generatione et Corruptione* (*On Generation and Corruption*) and *Meteorologica* (*Meteorology*, Books I and IV). Among the commentaries would have been *Tractatus de Sphaera* (*The Sphere of the World*) written around 1230 by Johannes de Sacrobosco, which provided a readable summary of the astronomy found in Aristotle and later Greek philosophers, such as Ptolemy. Sacrobosco's book profoundly influenced medieval art and literature, including probably the greatest poem of the Middle Ages, Dante's *The Divine Comedy*.

Dante wrote *The Divine Comedy* between 1308 and 1320, while William was studying in London. It is filled with elements drawn from *Tractatus de Sphaera* together with components from other medieval scholars, such as Roger Bacon and Robert Grosseteste,[5] texts that William would also have studied; plus a good deal from the poet's fecund imagination. Although it is largely fanciful, it illustrates how entangled theology and scientia were in medieval philosophy[6] and is therefore a great place to begin our exploration of the role of Occam's razor in the progress of science.

In his epic poem, Dante takes us on a tour of regions of the medieval universe. He begins his journey on earth from where he descends into hell and from there visits purgatory.* He finally ascends into heaven, accompanied by the spirit of Beatrice, his childhood sweetheart. Beatrice then takes Dante on a tour of ten heavens where they visit the orbs of the sun, the moon (quoted at the start of this section) and the planets Mercury, Venus, Mars, Jupiter and Saturn. The 'diamond-bright' material in the passage is a rotating sphere made of transparent crystal on which the moon (the 'eternal pearl') was thought to be moored. The rotations of the lunar sphere transported the moon around the earth on its monthly circuit. The sun, and each of the planets, were similarly propelled through their geocentric orbits by crystal spheres. It is on the lowest, lunar sphere, where Dante first encounters some of the supernatural inhabitants of the heavens, the 'faces' of the souls of the blessed.

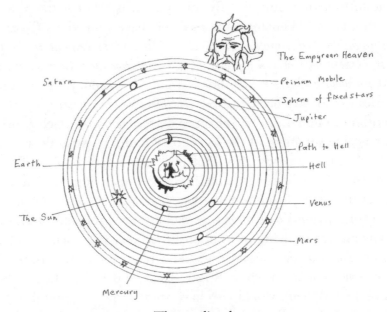

FIGURE 3: The medieval cosmos.

* In Catholic doctrine, a place where the souls of sinners who have managed to avoid hell must atone by a period of suffering before they can then ascend into heaven.

Dante's heaven is clearly a physical space; but is it science or theology? It is both. Souls and angels are there in abundance, but then so are questions that we would describe today as scientific. For example, during their tour, Dante and Beatrice embark on a lengthy discussion of the possible causes of the dark spots on the moon. This was a hotly debated topic among scholars in the ancient and medieval world because the moon, being an inhabitant of heaven, was expected to be unblemished. Some claimed that the spots were the stains of mankind's sins; whereas Beatrice discusses and dismisses another possibility, that the moon might have transparent regions. Science and theology are both inhabitants of the scientia of the medieval cosmos.

Dante continued his ascent, travelling through the five planetary spheres before visiting the celestial sphere that carried the fixed stars on their daily orbit. There was considerable discussion on the nature of the stars, whether they were, for example, spherical bodies attached to their sphere, or perhaps pinpricks in the heavenly sphere through which divine starlight shines. Beyond the celestial sphere was the highest heaven or *primum mobile* whose purpose, Beatrice explains, is solely to provide propulsion for the inner spheres carrying the stars and orbs. Beyond it is the dwelling place of God and the saints.

I should point out that Sacrobosco's astronomy text does not include angels or mention any other explicit theology, as it was based on much of Aristotle's almost secular astronomy. Nevertheless, the majority of people who studied Aristotle in the medieval world were theologians who sought ways to incorporate his astronomy into their notion of a Christian heaven, as reflected in their commentaries. Dante's poem thereby provides a glimpse of the heaven that William studied but also what educated men and women thought they were looking at when they glanced up at the night sky. Very far from our modern notion of a night sky filled with spheres of rock or fiery gas separated by vast voids, the medieval person saw the walls of heaven decorated with the sun, moon and stars. If they could, like Dante, rise into its heights and peel back the starry firmament, then they would have expected to see, along with the angels and saints, the face of God.

The medieval universe was thereby a peculiar amalgam of Greek astronomy and Christian theology. The theological components had been cobbled together from the Hebrew Bible and the writings of Christian theologians. To discover the origins of its scientific parts, we need to head both east from Newgate and backwards in time to ancient Mesopotamia.

## The orbs

Look up at the night sky on a clear night and you will see about two thousand stars. You may also see the moon and up to five visible planets. The moon is easy to spot. But which of the two thousand or so stars are planets?

An ancient Babylonian (1800–600 BCE) could have provided you with an answer. Their hot summer nights spent sleeping on cool rooftops made them very familiar with the movements visible in the night sky. They grew up recognising the constellations of two thousand or so *fixed stars* that twinkle as well as rotate in perfect circles around a point in the night sky marked by the Pole Star. But they also spotted five stars that didn't twinkle, nor follow circular paths, preferring instead to wander through a broad swathe of constellations known as the Zodiac. Their roaming habits earned them the name of wandering stars, or *planetes* in Greek.

The feature that most intrigued the ancient astronomers was the motion of the planets. Like most ancient people they drew a distinction between inanimate and animate objects. They believed that the inanimate kind tend to lie stationary unless given a push; whereas animate objects possessed the power of autonomous motion conferred by a supernatural soul that animated flesh. Since the heavenly bodies moved erratically across the sky without any visible mover, the Babylonians, along with nearly all ancient people, believed that they, like us, are animated by supernatural agents or souls. The planet we call Mercury was pulled around the sky by the chariot of the god Nabu. Similarly, Ishtar, Nergal, Marduk and Ninurta steered the planets we know today as Venus, Mars, Jupiter and Saturn. To provide the moon and the sun with their own independent motive forces, the Babylonians had them harnessed to the chariots of the sun god Sin and moon god Shamash.* The five visible planets plus the sun and the moon gave them, and us, the seven days of the week. Pinning a god to each fixed star would have been expensive in deities so the Babylonians opted for a simpler solution by attaching them to the inner surface of a hemispherical cosmological oyster shell that rotated from east to west around the Pole Star every day.

This deity-filled cosmos sounds quaint today but, in the absence of any understanding of gravity, the gods did the job in the heavens. As we shall discover, science is not about finding any kind of ultimate truth, it is about building hypotheses or models that we use to make useful predictions. The Babylonians' god-filled model of the heavens worked sufficiently well for its principal purpose of providing them with a calendar that their astronomers and astrologers used to predict the best time to plant, harvest, marry or make war.

## The spheres

Babylon fell to the Achaemenid Persian Empire in 539 BCE, but its astronomy survived and crossed the Aegean to be picked up by ancient Greek astronomers. There, the heavenly pantheon of Babylonian gods was supplanted by Greek deities, such as Aphrodite or Ares. However, the more philosophically minded Greeks, such as Anaximenes (585–528 BCE) of Miletus (a Greek town on the Anatolian coast), made the gods redundant, at least in the heavens, by replacing their divine drive with a concentric series of mechanical spheres whose rotations propelled the moon, sun, planets and stars around the earth and across the sky. To account for the obvious problem that no one could see the spheres, Anaximenes adopted

---

* Most ancient people did not differentiate between the five visible planets that we know today and the sun and moon – they were all 'planets'.

the approach that bedevilled pre-modern science: he invented an entity to fill the explanatory gap. He proposed that the heavenly spheres were made of a perfectly transparent, crystal-like heavenly element known as aether, or the fifth element, quintessence, from which we obtain the modern English word, quintessential.

There was, of course, no evidence for either spheres or aether; but, in the ancient world, they were an economic means of accounting for the heavenly motions as they replaced a pantheon of gods with just two entities. However, their presumed existence inspired mystics, philosophers, astrologers and astronomers to invent additional entities for millennia. Pythagoras (c.570–c.495 BCE), who was born on the island of Samos, claimed that the rotation of the spheres created a heavenly music audible only to highly tuned ears. A thousand years after Anaximenes, alchemists were claiming to extract pure quintessence from their potions, while two thousand years after Pythagoras, composers were still writing *music of the spheres*. Entities may become superfluous but they are often remarkably durable.

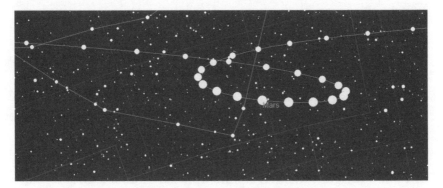

FIGURE 4: Position of Mars against background stars
on consecutive nights.

Yet, although crystal spheres worked well for the sun, moon or fixed stars that moved in perfect circles across the sky every day, they hit a major obstacle when attempting to account for the motion of those wandering planets. Not only were their paths non-circular but, as well as wheeling from east to west along with the fixed stars, they frequently changed course, in what we call today retrograde motion, to move from west to east. This had not been a problem for the ancient Babylonian planets driven by capricious gods, but how do you make an object on the surface of a rotating sphere wander?

The greatest philosopher of the ancient world thought he knew the answer. Plato was born around 428 BCE to a wealthy Athenian family. He became a pupil of Socrates and, after the elder philosopher's execution, founded the world's first school of philosophy, the famous Academy in Athens. There he lectured and wrote extensively on philosophy, the arts, politics, ethics and science, particularly Pythagorean mathematics and astronomy. His most influential idea, and one that was to profoundly shape the course of Western culture, was his concept of Forms* and the accompanying philosophical tradition known as philosophical realism.

Plato's realism encompasses all aspects of experience but is most easily explained by considering the nature of mathematical and geometrical objects such as circles. He asked the question, what is a circle? You might indicate a particular example etched into stone or drawn in the sand, but Plato would point out that, if you looked closely enough, you would see that neither it, nor indeed any physical circle, was perfect. They all possessed kinks or other imperfections, and all were subject to change and decayed with time. So how can we talk about circles if they do not actually exist?

The problem is not restricted to geometrical objects but is apparent in every word we give to a class of objects or concepts, for example, rocks, sand, cats, fish, love, justice, law, nobility, and so on. Individual examples or instances are different from each other and none corresponds to the idealised cat, rock or noble; yet we have no trouble in recognising and talking about them. So what are we comparing them against to identify them as circles, rocks, fish or cats?

Plato's extraordinary answer was that the world we see is a pale reflection of a deeper reality of Forms, or *universals*, where perfect cats chase perfect mice in perfect circles around perfect rocks watched over by perfect nobles. Plato believed that the Forms or

* *Form* tends to be capitalised when referring to the philosophical entity.

universals are the true reality that exists in an invisible but perfect realm beyond our senses. His system is often known as *philosophical realism* to denote that Plato, and his followers, believed that Forms or universals are not only real but are the ultimate reality that give rise to our sensory perceptions.*

Plato graphically illustrated his model in his famous Allegory of the Cave in which he compared the human experience to that of people chained facing the wall of a cave lit by a fire. Real objects (analogous to his Forms) pass between them and the fire but the inhabitants of the cave can only perceive their own shadows projected onto the cave wall. They are convinced that these shadows are the real world and are completely unaware that there is another, more vibrant reality, if they could only turn around to face it. Plato similarly insisted the real world of the Forms cannot be perceived by our senses but only our minds. He advised that the philosopher's mind, 'along with the whole soul, must be wheeled round from that which is subject to becoming [the visible world of our experience] until it is able to endure the contemplation of that which is, and the most resplendent part thereof: and this we declare is the Good.'[7]

Nobody is sure where Plato located his realm of perfect Forms but, in his *Phaedrus*, they are in a 'place beyond heaven'. As cohabitants of the heavenly realm, the planets must be perfect in every way, including travelling along geometrically perfect circular paths and at uniform speed. That this claim was plainly contradicted by his senses must, Plato insisted, be a consequence of the inferior vantage point available to humanity from within the terrestrial cave of our senses. He urged his followers to ignore the distorted lens of their senses and instead use their intellect to discover the 'circular motions, uniform [speed across the sky] and perfectly regular, [that] are to be admitted as hypotheses so that it might be possible to save the appearance presented by the planets'.[8] Thus, *saving the appearance of the planets*, as this quest came to be called, became the primary mission of astronomers for more than two thousand years.

* Not to be confused with being realistic in the sense of sensible and practical.

The first to take up the challenge of saving the appearance of the heavens was Plato's pupil, Eudoxus of Cnidus (410–347 BCE) who, in a pattern that will become familiar, added more spheres. Imagine you are standing in Plato's cave, which lies at the centre of a simplified version of Eudoxus's model consisting of just a single sphere, which we will represent in Figure 5 as a band-like section of the crystal sphere (though remembering that Eudoxus imagined an entire sphere). Attached somewhere on the inside of the band's circumference is a bright light, which we will call the 'Planet'. Now imagine watching only the light as the band rotates, and what you will see is that the Planet follows a perfectly circular path. Further imagine fitting a complete crystal sphere to the inside of the band so that the band and sphere are concentric. The band now runs on rollers so that it rotates smoothly along a fixed track over the surface of the crystal sphere. From your perspective at the centre of both band and sphere, the planet again follows a circular path. But now, as the band rotates, we allow the inner sphere to also rotate on yet another axis. The planet's motion remains perfectly circular, seen from its own perspective, yet from your perspective in the cave, it now appears to follow a more complex path, which is the superposition of two circular motions. It moves like the planets in the sky.

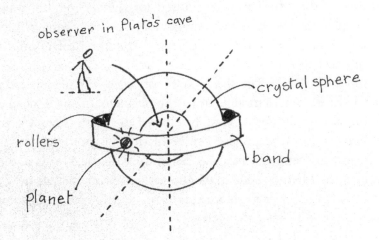

FIGURE 5: View of the planetary motions in Eudoxus's model.

Eudoxus's model worked well enough but it needed twenty-seven spheres. Plato's mechanically minded pupil Aristotle added more spheres to act rather like modern ball-bearings to prevent motion from one sphere being transmitted to adjacent spheres. The number of celestial spheres jumped to fifty-six. Still, a problem remained. No number of rigid rotating spheres could accommodate another feature of planetary motion – that the planets regularly wax and wane in luminosity. To maintain their constancy of illumination, they would have had to, alternately, move closer (wax) and further away (wane) from the earth. How could they perform this manoeuvre on the surface of a rigid sphere?

A solution was devised by the last great astronomer of antiquity, Claudius Ptolemaeus (Ptolemy, around 90–168 CE) who lived in the Graeco-Roman city of Alexandria famous for its Great Library. His first move was to incorporate an idea from the third-century BCE Greek astronomer Apollonius.* Imagine that instead of pinning the imaginary planet in Figure 5 to the outer band, we hang it instead from a small rotating wheel, rather like a seat hanging from a Ferris wheel, whose hub is attached to the band. The sphere and ring rotate the planet exactly as before but now the Ferris wheel's rotation delivers an additional epicycle that moves the planet alternately nearer and further from your vantage point. The addition of the wheel now accounted for the waxing and waning of the planets but how can any kind of wheel swing a planet through a supposedly rock-solid crystal sphere? Ptolemy did not attempt to explain.

Even with all this complexity, the motions of the planets did not entirely fit Ptolemy's model. To fix the problem, he introduced two additional complications. First, he shifted the earth (Plato's cave in the figure) from the precise centre of the sphere's rotations to a point, just off-centre, which was called the *eccentric*. He also quietly dropped the Platonic principle of uniform motion by allowing each planet only to appear to rotate at uniform speed from an imaginary point in space termed the *equant*.

* Apollonius was actually born in Anatolia.

Ptolemy described his final geometric model of the cosmos in his *Almagest*, written in about 150 CE. It was extremely complex, including around eighty circles, epicycles, eccentrics and equant points. It was also profoundly non-physical as it involved the planets, on their heavenly Ferris wheels, rotating clean through the supposedly solid crystal spheres. It was also geocentric, with the earth, rather than the sun, at its centre. Yet, astronomical predictions made using the *Almagest* model of the cosmos were pretty accurate, accounting for much of the observed motions in the heavens as well as the date of events, such as eclipses, so much so that it became the last word in astronomy for over a thousand years. It was widely studied in the Arabian world and most of the astronomy of Johannes de Sacrobosco, in *Tractatus de Sphaera*, which William of Occam probably studied at Oxford, came via the Arabic translations from the *Almagest*.

How can a model that is so wrong get so much right? This is actually a very deep question that challenges the widespread notion that the mission of science is to peer beyond the limitations of our senses and unguided intellect to discover what the world is *really like*. If scientific models, such as Ptolemy's, that are based on so many wrong assumptions, can still make accurate predictions, then how can we tell whether a particular theory or hypothesis is right or wrong? Perhaps today's scientific models that similarly account for most of today's data are just as flawed as Ptolemy's? How then do we discover the truth?

As you might guess, the answer to this conundrum involves Occam's razor, but its adoption will entail abandoning what might be called the naive view of science as a quest for the truth in favour of a more nuanced and perhaps unsettling acceptance that the *truth* will always be beyond our grasp. Nevertheless, despite this limitation, and with Occam's razor in hand, science can, and does, help us to make sense of our world such that we can fly rockets to remote planets or free billions from the yoke of disease or starvation. Science may not know where it's going, but the journey is amazing.

## *The fall of the heavens*

Ptolemy's model was the last great achievement of classical science. His home city of Alexandria continued as a centre of learning well into the Christian era. Its great library was so famous that, in the first two centuries CE, Alexandria was regarded as the capital of learning throughout the ancient world. The Alexandrian Museum, established around 300 BCE, is arguably one of the world's first universities that counted eminent scholars, such as Euclid, among its faculty. Its last director was a mathematician called Theon of Alexandria. Theon's learned and beautiful daughter Hypatia had achieved great fame in her own right as a mathematician, philosopher and teacher, and as such epitomised the ideals of Hellenic culture. She is the first female mathematician of whom we have certain knowledge.[9] She continued to teach and worship the pagan gods even after Emperor Theodosius's edict against the old Greek religion. John, Bishop of Nikiû, takes up the story of her fate in 415 CE when 'A multitude of believers in God . . . dragged her along till they brought her to the great church . . . And they tore off her clothing and dragged her through the streets of the city till she died . . . and they burned her body with fire.'[10] St Jerome, the translator of the Vulgate Bible, wrote that 'the stupid wisdom of the philosophers' had been defeated. The crystal spheres that had provided 'the course of the stars' for the Greeks and Romans were shattered and the model of the cosmos reverted to a flat earth surrounded by a Hebrew tent to hold up the stars. Severian, Bishop of Gabala, in about 400 CE, wrote in his Six Orations on the Creation of the World, 'The world is not a sphere but a tent or tabernacle.'[11]

# 2

# The Physics of God

The period of Western European history that used to be known as the 'Dark Ages'* saw the population of the region plummet from about 9 million in 500 CE to about 5 million, four hundred years later. Literacy levels dropped dramatically and construction of monumental architecture almost ceased. The period saw massive migrations as successive waves of invaders rushed to fill the vacuum left by the collapse of the Roman Empire, and refugees sought to escape the subsequent chaos.

Some learning did however survive in Europe, often on the fringes of the former Empire, such as in Northumbria and Ireland in the British Isles. Scholars from both these regions, such as Alcuin of York (735–804) and John Scotus (815–77), later travelled across Europe to contribute to the Carolingian Renaissance of the eighth and ninth centuries that eventually settled into what tends to be known today as the Early Middle Ages or Early Medieval period.[1]

The Carolingian Renaissance brought technological innovation that included the introduction of the heavy plough, the stirrup and the windmill. Yet, although these innovations led to some progress, it was against a background of stasis or only incremental change. We don't have reliable productivity data for the whole of the Early Middle Ages but a pattern can be seen in, for example,

---

* The term is hardly used by today's scholars who insist that these centuries were not nearly as dark as their reputation. However, I believe there is some justification for use of the term in the chaotic years that immediately followed the fall of Rome in the West, if only because its loss of literacy made the period 'dark' from our perspective.

the agricultural productivity of England from 1200 to 1500,[2] which shows extremely shallow growth over three hundred years. This kind of virtual stagnation – what we might today call linear growth – had also been typical of earlier civilisations, such as those of ancient Babylon, Greece or Rome, as well as in pre-industrial China, India and Mesoamerica. In fact, this pattern of linear growth interrupted by occasional spurts of progress seems to have been characteristic of nearly all human history, except for the last few hundred years which have exhibited a pattern of exponential, or accelerating, growth until very recently. We will return to the question of how and why human progress shifted from a linear to an exponential gear in later chapters; but, as you might expect, I believe Occam's razor played a key role.

After the fall of Rome and the the establishment of the eastern Byzantine Empire centred on Constantinople, western Europe none-theless retained Roman Latin as its lingua franca. This allowed, for example, Roman law to be adopted throughout most of the West. Science and philosophy, however, were nearly always written in Greek, with the result that much of that learning was lost to the West. And while the Greek-speaking Byzantine Empire enjoyed continued access to the ancient Greek texts, for reasons that remain unclear, the Byzantines seemed to have taken little interest in Greek science.

A handful of Greek texts had, however, been translated into Latin before the fall of Rome. One of the most famous was written by the Christian Roman aristocrat Boethius (around 475–525 CE). Entitled *The Consolation of Philosophy*, Boethius wrote it in his jail cell while awaiting a horrible execution on a charge of treason. It featured his imagined dialogue with Lady Philosophy about the merits of philosophy, particularly Plato's. It became immensely popular in the Middle Ages and was read by nearly all the literate minority. It remains in print today.

Latin translations of fragments of Plato's dialogues also made it into the West, including much of his *Timaeus*, which went on to profoundly influence the intellectual development of Augustine of Hippo (later sanctified and known as St Augustine). In his *The City of God*, which became extraordinarily influential in the Middle Ages, he describes how God 'brought in my way . . . some books of the Platonists translated from Greek into Latin'. The effect was so overwhelming that, he tells us, '[I] entered into my own depth and was thus set on the road to conversion'.

Despite its popularity, *The City of God* projected a bleak view of humanity. Augustine wrote it after the sack of Rome by the Visigoths in 410 CE, and the long list of barbaric cruelties perpetrated over three long days of murder, rape and looting probably coloured his view of humanity as 'a heap of depravity'. The chaos may also have inspired Augustine's adoption of philosophical realism as a means of reconciling the savagery with his vision of a benevolent Christian God. He co-opted Plato's world of Forms to claim that worldly imperfections were a pale and distorted reflection of the invisible but perfect realm of heaven.

In his *Confessions*, Augustine did ponder questions that today we would label as scientific, such as the nature of time; yet these were always framed within a theological context, such as how an unchanging God could do anything within time.[3] Augustine was also suspicious of the tendency of human intellect to stray beyond theology, warning that

> I mention another form of temptation . . . a certain vain desire and curiosity, not of taking delight in the body, but of making experiments with the body's aid and cloaked under the name of learning and knowledge . . . Certainly the theatres no longer attract me, nor do I care to know the course of the stars . . . What concern is it of mine whether heaven is like a sphere and the earth is enclosed by it and suspended in the middle of the universe, or whether heaven like a disk above the earth covers it over on one side?[4]

The consequence of Augustine's disdain for 'another form of temptation' was that science, like the economy, stagnated in Europe's Early Middle Ages.

## *The earth becomes round again*

Fortunately, St Augustine's influence did not extend into the Middle East and most of the Christian zealots who had usurped the old Roman religion were driven out by the Arab conquests in the seventh century. The region's Islamic rulers were far more tolerant of ancient learning than those in the West, and intellectual centres such as Baghdad's House of Wisdom, founded by Caliph Al-Mansur in the eighth century, sprang up across the Islamic world. Fragments of Greek manuscripts rescued from ancient libraries such as that at Alexandria were highly prized. The rescued works of Plato, Aristotle, Pythagoras, Euclid, Galen and Ptolemy were eagerly translated into Arabic and commented on by Greek-reading Islamic scholars, such as Al-Kindi (born around 801) in Baghdad who wrote influential commentaries on Aristotelian logic. Born in what is now Aleppo in northern Syria, the tenth-century female astronomer Mariam al-Asturlabiyy was renowned for the making of astrolabes. Greek science was not only studied but expanded by many Arabic-speaking scholars, such as the Basra-born Ibn al-Haytham (965–1040 CE) whose seven-volume treatise on optics, *Kitab al-Manazir* (*Book of Optics*), described his ground-breaking experiments on reflection that demonstrated, for example, that light always travels in straight lines. He was also the first to recognise that sight requires light to enter the eye. Islamic dominance in mathematics through the Early Middle Ages is reflected in English words with Arabic roots, such as algebra and algorithms while words like alchemy, alcohol and alkali are witness to Islamic innovations in chemistry. Arabian technological innovations, for example windmills, distillation, pens and buttons, had been unknown in the ancient world.[5]

The West remained an intellectual backwater until the ascent to the papacy of Gerbert of Aurillac, a classical scholar, geometer, astronomer and philosopher, who became Pope Sylvester II in 999 CE. Gerbert had travelled widely before becoming Pope, including to Spain, where he encountered Arabic and Greek manuscripts. He encouraged a rediscovery of, and respect for, Greek and Arabian science as well as introducing Hindu-Arabic numerals. He even owned an armillary sphere, a model of the heavens built from concentric metal rings centred on a spherical, not flat, earth. Contrary to the myth, no educated person in the Middle Ages thought the earth to be flat.

This trickle of learning from the ancient world became a flood after the Moorish kingdoms of Iberia and Sicily fell to the Christian Reconquista in the late twelfth and thirteenth centuries. The crusading knights who broke open the doors of the great Islamic libraries of Toledo, Córdoba and Palermo found the most astonishing treasure of all: their own forgotten past. An intellectually starved Europe came to realise that Greek and Roman philosophy and science, a world of ideas that they had believed irrevocably lost, had instead been preserved, and extended, in the books of their enemy. It was one of the most ironic volte-faces in history. Islamic scholars, such as al-Kindi and the Persian polymath Ibn Sina (born 980 in Hamedan, Iran) known as Avicenna in the West, or Ibn Rushd (born 1126 in Córdoba, Spain and known in the West as Averroes), had laboured for centuries translating Greek works written by the greatest minds of ancient Europe into Arabic. European scholars who read Arabic consumed these works and translated them into Latin.

These translations of philosophy and science ignited a blaze of intellectual activity to initiate an unprecedented period of learning in the West that is sometimes known as the twelfth-century Renaissance. Cathedral clerical schools that had been set up across Europe during the Carolingian era opened their classes to Greek and Arabic texts. When King Louis IX of France heard that a Saracen sultan had established a library with a vast collection of books, he decided to do the same for the Parisian college founded by Robert de Sorbon around 1150. The Sorbonne, as it came to be known, became the nucleus of the University of Paris, described by the scholar and poet Jean Gerson as 'the Paradise of the world, where is the tree of knowledge of good and evil'.

The rediscovered philosopher who had, by far, the biggest impact in the Late Middle Ages was Aristotle. When Latin translations of

Arabic sources arrived in the West, scholars – or the scholastics, as the Aristotelian scholars came to be known – pounced on the works of Aristotle and his Arabic commentaries as if they were rediscovered treasure – which of course they were. Robert Grosseteste (1175–1253), who became Bishop of Lincoln, translated many of Aristotle's works while studying at the University of Oxford, and between 1220 and 1235 wrote a host of scientific treatises on philosophy, astronomy, optics and mathematical reasoning. His fellow Oxford scholar, the Franciscan Roger Bacon (1219–92), translated Aristotle's treatises on 'The Science of Perspective' and 'On Experimental Knowledge', helping to revive interest in experimentation. In Paris, another translator, Albert the Great (1200–80), wrote his *Commentary on Aristotle's Physics* as well as his own *De Mineralibus* (*Treatise on Minerals*) in which he fused Aristotle's theory of causes with his own observations and even experiments, essentially founding the modern science of mineralogy. In it, he insisted that 'The aim of natural philosophy is not simply to accept the statements of others, but to investigate the causes that are at work in nature.'

European scholastics not only imported Greek and Arab learning into the West, but they also applied it to new areas of study. For example, in his *De Colore* published around 1225, Robert Grosseteste described a three-dimensional geometric colour space that isn't so different from how we describe perceived colour today. He was also the first to point out that the rainbow is a result of refraction.[6] In his *Opus Majus* written around 1266, Roger Bacon imported much of Aristotle's natural science, grammar, philosophy, logic, mathematics, physics and optics but added a study of lenses that may have inspired the development of spectacles.

While the revival of science in Europe was a huge breakthrough and involved some genuinely new ideas as described above, many of its advances simply brought European scholars up to speed with the ancient world and advances further east. Much of Robert Grosseteste's work on optics was based on Al-Kindi's *Optics*; whereas Bacon's 840-page treatise *Opus Majus* was mostly drawn from Ibn al-Haytham's *Book of Optics*. Even terms such as 'experiment' in Bacon's works are misleading to modern ears, as the medieval term meant merely observations from experience, such as observing the colours of a rainbow, the boiling of water or the attraction of a magnet. Bacon also gave the first account of gunpowder and its use in fireworks in the West, but it probably originated in the Islamic world. However, the most important distinction between the medieval scientia that Grosseteste and Bacon studied and modern science was that they both considered they were investigating a branch of theology. For example, Grosseteste believed that all light was an emanation from God[7] and both he and Bacon insisted that theology was the foundation of all sciences.[8]

Despite its subordination to theology, the importation of 'pagan' ideas into Christianity was not uniformly welcomed by theologians. Many traditionalists feared that reading Aristotle was leading young scholars into heresy. The matter came to a head on 7 March 1277, when the Bishop of Paris, Stephen Tempier, issued a series of prohibitions that banned the teaching of 219 philosophical and theological theses, mostly from Aristotle. Many prohibitions were addressed at theologians who dared to place Aristotle's logic above God's omnipotence, for example, arguing over whether God could make a vacuum given that Aristotle had insisted a vacuum was logically impossible. The 1277 prohibitions only strictly applied in Paris, yet their influence led to a more critical attitude to Aristotle across most of Europe's leading universities.

Today we know that this setback was only temporary. After a period of retrenchment, progress continued, the prohibitions were forgotten and Aristotle came once again to dominate the curriculum in Western universities. However, this happy outcome was not guaranteed. Two hundred years earlier a similar anti-Hellenistic and anti-rationalist backlash in the Islamic world, promoted by the Ash'ari theological school of Sunni Islam, had snuffed out the 'Golden Age' of Islamic science. Thereafter, Arab scholars restricted their studies to the 'literal truth' of the Koran.[9] That medieval European science was not similarly strangled at birth owes much to the influence of the greatest theologian of the medieval age, who had arrived in Paris thirty years before the prohibitions, Thomas Aquinas (1225–74).

## The Dumb Ox

Born in 1225 into a wealthy Italian family in Roccasecca, central Italy, Aquinas was the ninth child of Theodora Carraciola, Countess of Teano. Thomas received his primary education at a studium generale in Naples where he first encountered the ideas of Aristotle and his Arab commentators, particularly Ibn Rushd (Averroes) and the Sephardic Jewish philosopher Moses ben Maimon (born in 1138 in Córdoba, Spain) known as Maimonides.

Thomas's family expected him to become a Benedictine abbot, a lucrative position that could earn the family new lands. Thomas had other ideas. He wanted instead to join the Dominicans who, like Occam's Franciscans, were a mendicant religious order known for their respect of the new learning. To his family, this was the medieval equivalent of joining a cult, as the mendicants were considered hardly more respectable than wandering beggars. To prevent Thomas taking up this humble vocation, his family resorted to locking him up in the family castle. His brothers even smuggled in a prostitute in a ploy to tempt him away from the saintly life. Thomas is said to have chased her away with a burning stick. His sister eventually helped him to escape by lowering him in a basket from the castle tower into the waiting hands of Dominican co-conspirators. Aquinas escaped Italy and travelled to the centre of learning in medieval Europe, the University of Paris, arriving there about 1245.

By this time, the most prolific and influential translator of Aristotle in the West, Albert the Great, had been teaching in Paris for five years. The novice friar came to the notice of the eminent theologian as a shy retiring new student constantly mocked by his fellows as the 'dumb ox' for his heavy build and premature baldness. Yet Albert recognised the brilliance of the young scholar and predicted that 'the bellowing of this ox would be heard throughout the world'. He was right.

When Albert moved to Cologne to teach at its studium generale, Aquinas followed him a few years later, then returned to Paris to study for his master's in theology and write a commentary on *The Four Books of Sentences*.[10] This text had been written in the previous century by the French scholar Peter Lombard as a collection of essays addressing thorny problems that theologians continue to ponder, such as 'What is free will?' as well as questions that are a kind of fusion between theology and science, such as 'In what manner can waters be above the sky, and what kind are they?' These questions reflect a key aspect of the medieval worldview, that is also apparent in Dante's great poem, the belief in a single realm that includes both natural and supernatural elements. After outlining the question, each chapter included a collection of responses written by the Church Fathers. All theology students in the Middle Ages were required to write an extensive commentary on *The Four Books of Sentences*, rather like a modern doctoral thesis.

In 1259 Aquinas moved back to Italy and, sometime between 1265 and 1274 (just before the Paris condemnation of Aristotle's ideas), he wrote his most important work, *Summa Theologica*, which became so influential that it succeeded in almost canonising Aristotle within Western Christianity. Indeed, in *The Divine Comedy*, Dante and Beatrice meet Peter Lombard, Albert the Great and Thomas Aquinas in the sphere of the sun; but it is only Aristotle who is known as 'The master of those who know'.

Aquinas's aim was to build a new rational model of the cosmos based on Aristotle's science but incorporating God as well as angels, saints and demons. However, to make the Christian God consistent with Aristotelian science, Aquinas first had to prove His existence. To accomplish this feat, Aquinas appropriated the Greek philosopher's analysis of change and motion. Aristotle had written, 'everything that is moved is moved by something'. However, in contrast to our modern tendency to look for single causes for events, such as a spark being the cause of a fire, Aristotle provided events with four different causes of change: material, formal, efficient and final. So, for example, bricks might be the material cause of a house, whereas its shape or plan would be its formal (as in 'form') cause, its builder would be its efficient cause, and its final cause or *telos* would be a place for human habitation.

The first three causes still make sense, although we might quibble with whether they need to be distinguished at all; but Aristotle's fourth cause, the telos, is very different from anything in modern science because it reverses the usual temporal order between agency and outcome. Whereas bricks, plan and builder precede the house, Aristotle's final cause lies in the future. Yet, to Aristotle and Aquinas, the telos remained as much a cause of the house's construction as its bricks. This makes some sense for human artefacts such as houses, but Aristotle believed that everything that happens does so for a final purpose. So, rocks fall to earth because the telos of a rock is to be as close as it can get to the centre of the earth; whereas the telos of the moon is to orbit the earth in perfect circles. Extended into the living world, the telos of lower beings, such as pigs, was to serve the higher ones, such as humans, by being eaten. The Roman philosopher Vero even went so far as to insist that the telos of life for a pig was to keep its meat fresh.

But where do you stop? After all, such a hierarchy of *teloi* has the potential to lead to an infinite regress: the telos of a turnip is a hungry pig and the telos of a pig is a hungry human and so on. Aristotle avoided this problem by capping his hierarchical chains of causes with the prime mover, or God, who was the first and the final cause of everything. This, for Aquinas, clinched the deal he struck between theology and Aristotle's philosophy. Although Aristotle's prime mover was a rather remote impersonal entity, more an 'it' than a 'him' or a 'her', and quite unlike the personalised Christian God, Aquinas enthusiastically incorporated Him (God was always male to medieval theologians) into theology so that the God of the Bible became both the efficient and final cause of everything and everyone in the medieval world.

This importation of Aristotle's four causes into Christian philosophy provided Aquinas with the huge windfall of four of his five scientific 'proofs' of God. Three of his Five Ways or proofs insisted that the Christian God must have been the material, formal and efficient first cause of all objects and events in the world. He applied the same logic to his fourth proof, arguing that 'some intelligent being exists', he wrote, 'by whom all natural things are directed to their end; and this being we call God'. God was thereby the final cause, the telos, of everything that ever happened or ever will happen. Aquinas's fifth proof, which is often known as the 'argument from degree', was a variant of the famous ontological* argument put forward a century earlier by the French philosopher Anselm of Canterbury (1033–1109). Aquinas argued that any ascending order of existing things must be capped by the greatest being, which can only be God.

By providing five proofs of God, Aquinas claimed to have successfully incorporated the Christian God into his model of the cosmos based on Aristotle's science. He had, he claimed, proved that theology was a science, indeed the 'Queen of Sciences'. Yet there was more. Aquinas next attempted to show that his theological scientia could explain even miracles.

## The taste of God

Aquinas's final philosophical sleight of hand would, a generation later, earn William of Occam his charge of heresy and, a century or so later, spark the great schism of Western Christianity. It concerned the miracle of the Eucharist, or communion sacrament, that lies at the theological core of the Christian Mass. In it, the priest calls on God to, quite literally, transform bread and wine into the body and blood of Christ. Most theologians would have categorised this miracle of transubstantiation, as it is called, in the same class of events as Jesus turning water into wine or Moses parting the waters of the Red Sea. It involves intervention of the divine and was thereby not expected to conform to the usual rules that apply to normal life. Aquinas, however, was convinced that he could incorporate even miracles into his scientific model of the world. To do so, he reached for another gift from the ancient world, philosophical realism.

* Ontology is the branch of philosophy that deals with questions about what does and does not exist. It contrasts with epistemology which is about what we can know.

St Augustine had already imported Plato's Forms into the early medieval Church where they became ideas in the mind of God. However, by the thirteenth century, they had been supplanted by Aristotle's universals. These were similar to Platonic Forms but they existed out there in the world, filling each object with a kind of essence of what it was. Each circular object shared in the perfect essence of circularity; all nobles possessed their share of nobility; all fathers partook in the essence, or universal, of fatherhood.

From a scientific perspective, Aristotle's universals were a slight improvement on Plato's Forms since they were considered to exist in the world rather than in some invisible realm. Yet this created a problem: how do we gain knowledge of them? Vast reams of scholarly parchment were dedicated to answering this question without ever coming to a conclusion. An additional problem was that there were so many of them: at least one for each noun or verb in our language. In his 'Categories', Aristotle tried to bring some order into this explosion by placing each universal into one of ten categories. These included substance, quantity, quality, place, relation, position and so on.

Exactly what Aristotle meant by his categorisation of universals is still debated but we might equate the first of them, substance, with the modern concept of matter. It represents the unchangeable *essence* of an object, its composition from the elements of earth, air, fire or water. The other universal categories, often known as *accidents*, were attached to an object's substance, providing it with its characteristic look, feel, taste, shape and smell. For example, all round objects possessed the universal of roundness and if, like cherries, they naturally occurred in pairs, then they also possessed the universal of twoness as well as the universal of sweetness, all attached to the substance of cherry. This was how a cherry became a cherry. Whereas the substance of an object was considered fixed, its accidents, such as colour or shape, were constantly in flux as, for example, a cherry grows and changes from bright to dark red as it ripens.

Universals were central to medieval philosophy and science because they underpinned the foundation of its logic: the Aristotelian syllogism. The basic structure is often illustrated with reference to Plato's famous teacher: Socrates is a man, all men are mortal, so Socrates is mortal. The logic is founded on the principle that all objects can be classified according to their universals, such as the universal of *manness*. Once this is accepted, then so long as you know the universal's accidental properties, such as mortality, then you can make definitive scientific statements that are bound to be true.

Syllogistic logic works well for the example above but, if we make a small adjustment – Socrates is a man, all men have beards (as they might have done in ancient Greece), so Socrates has a beard – then it becomes less clear that it is a good way of reasoning about the world. As we will discover, William of Occam's undermining of syllogistic logic was one of the inspirations for his razor. Nevertheless, for Aquinas, universals were much more than a tool of logic. He believed that their ultimate reality resided in God's mind such that study of them provided insights into God's plan. They were nothing less than a shadow of heaven in the world, filling all earthly objects with the direct presence of the deity.

Aquinas realised that the miracle of the Eucharist was a thorny problem for universals because the bread's substance – its unchangeable essence (according to Aristotle) – was believed to change into a very different substance, flesh. This is why the miracle is called transubstantiation. So, after the miracle, the bread, while still looking like bread, was (and is for Catholics) believed to be composed of the substance of Jesus's flesh. The question that vexed the scholastics was what were the accidents of bread – its bready taste or crumbly appearance – doing attached to the flesh of Jesus?

In his *Summa Theologica*, Aquinas devised an ingenious solution. During the miracle, Aquinas claimed, the bread's accidents, its taste, texture, colour and so on, attached, not to its usual substrate, substance, but to its quantity, which remained unchanged by the miracle. There was one bread before the miracle and one Jesus afterwards. This allowed the taste of bread to remain, even after its substance had vanished. The miracle of transubstantiation was therefore, Aquinas claimed, entirely consistent with Aristotle's science: a pearl in the crown of the 'Queen of Sciences'.

Such was the enormous influence of Thomas Aquinas, who was canonised* only fifty years after his death, that this bizarre explanation became standard doctrine in the Christian Church where it remains so today in the Catholic Church.† As the medievalist scholar Edith Sylla wryly observed, Aquinas considered that 'he was not introducing an alien philosophical element into sacred doctrine, but using a purified reason that combined with revelation to make a single sacred science – the water of philosophy when mixed with the wine of revelation was turned into wine.'[11] For this finale of his Aristotelian scientia, Aquinas transubstantiated theology into science.

Nearly two millennia earlier, Socrates had insisted that he was wiser than another man because 'I do not fancy I know what I do not know.'[12] This was the fundamental problem of scholastic scientia.‡ The scholastics thought they knew everything but, in reality, they knew nothing. Their entity-bloated scientia could explain everything but, without a principle of parsimony, it predicted nothing.

Aquinas's *Summa Theologica* was never completed. In December 1273, the great theologian had a vision while saying Mass that left him unable to write or even dictate as, compared to his vision, 'All that I have written appears to be as so much straw.' A generation later, a scholar arrived in Oxford with a tool to cut through all that straw.

* Made a saint under Catholic doctrine.
† See, for example, http://www.faith.org.uk/article/a-match-made-in-heaven-the-doctrine-of-the-eucharist-and-aristotelian-metaphysics.
‡ Perhaps the closest modern analogy is the 'experimental theology' imagined by Philip Pullman in his *His Dark Materials* trilogy.

# 3

# The Razor

## William goes to university

William's progress through the *trivium* and *quadrivium* at Greyfriars in London probably took between three to six years. He must have impressed his teachers because he was then selected to study for a doctorate in theology. Greyfriars was loosely affiliated to Oxford, so sometime around 1310, when William was about twenty-three, he set off to continue his studies at England's first university to be trained as a medieval scholar or clerk.

Oxford was two days' ride north-west from London along a busy track. The route was regularly attacked by bands of robbers so novice students tended to group together to be escorted by a professional armed 'fetcher'. William probably joined such a group. We might imagine him as Geoffrey Chaucer's young clerk (the term derives from the Latin *clericus*, meaning cleric or clergyman) from *The Canterbury Tales* who 'had begun the study of logic' but who

> *would rather have at the head of his bed*
> *Twenty books, bound in black or red*
> *Of Aristotle and his philosophy*
> *Than rich robes, or a fiddle, or an elegant psaltery.**

After arriving in Oxford, William joined a Franciscan friary, possibly located in Greyfriars Hall in Iffley Road. The university had been established only a century or so earlier and was much

---

* Lute-like musical instrument.

smaller than its modern counterpart, consisting of a handful of colleges, including Balliol and Merton, plus several schools established by the Franciscan and Dominican orders. Most students were not friars or monks but they all had to be tonsured and dressed in clerical robes in order to enjoy clerical privileges. One of the most useful privileges was that, rather than being subject to secular courts, students accused of a crime would be tried by the ecclesiastical courts. These were presided over by the university chancellor and could, on occasion, allow an errant scholar to quite literally get away with murder.

Students from different regions of England, Scotland, Wales and Ireland, many as young as fifteen, formed gangs that regularly engaged in brawls. Skirmishes between seculars and the various holy orders were also common, as were 'town and gown' conflicts. Shortly before William's arrival, a dispute between the university and a group of friars had seen them excluded from university functions and their church attacked and desecrated by fellow students. Student conflicts regularly escalated to injury and even deaths. In 1298, a scholar named Fulk Neyrmit was killed by a townsman's arrow while leading a massed assault in the High Street armed with bows and arrows, swords, bucklers, slings and stones.[1] That same year, an Irish scholar, John Burel, was stabbed to death in a tavern brawl. Knife fights were a common source of injury as, throughout medieval England, pretty much everyone, including friars and monks, carried their own knife to table. The historian Hastings Rashdall remarked of Oxford that there are 'historical battlegrounds on which less blood has been spilt': an observation supported by a recent estimate of the homicide rate in fourteenth-century Oxford as far higher than that of the most violent cities today.[2]

Away from the melee, William would have attended lectures in his home friary and also in neighbouring friaries and university colleges. Once graduated, he was required to deliver lectures. These could either be standard university talks lasting about an hour, or 'disputations' in which students listened to masters debating contentious issues. Classes were held in rooms not dissimilar from those still to be found in the older colleges of Oxford or Cambridge, with wooden benches and desks for the students and a lectern for the master. However, unlike modern lecture theatres, the seating was unraked, so student and master inhabited the same space. This probably added to the generally boisterous atmosphere as many students, particularly the seculars who paid for their tuition, would often jeer and hurl insults at a master who was considered not to be earning his fee.

As a theology student, William's principal textbook would have been Lombard's *The Four Books of Sentences*. The question that particularly attracted his attention was 'Is theology a science?' Thomas Aquinas had insisted that theology was not only a science but was the 'Queen of Sciences'. William disagreed.

## Disputations

Sadly we don't have an image of the young William, as in the fourteenth century portraits were reserved for only the most powerful people. However, thanks to the scribblings of a young scholar in the margins of his notes, we do have a sketch of William, albeit twenty years after his undergraduate days. It was drawn by Conrad de Vipeth of Magdeburg, clearly a fan of the older scholar, who penned it in his own copy of Occam's *Summa Logicae*, during a visit to Munich. In Conrad's sketch, the slim, tonsured friar appears somewhat wistful and delicate.[3]

William completed his commentary on *The Four Books of Sentences* sometime between 1317 and 1319, when he was around thirty. Thereafter, he was required to lecture at Oxford, and possibly London. It was at this stage that his commentary came to be published. The common practice was that one of the students attending a lecture took detailed notes in ink on vellum, known as *reportatio*, which could be copied by other students both inside and outside of the university. The lecturer might correct and amend the student's account to release an approved copy, known as an *ordinatio*. William is known to have completed an *ordinatio* of his commentary on Lombard's first book of *The Four Books of Sentences* around

1320, but the only surviving copies of his commentaries on the other three volumes remain as *reportationes*. Disputations were also recorded and when corrected by the lecturer were called *quodlibets*. William completed seven of these between 1321 and 1324. Around this time, he also wrote a lengthy exposition on Aristotle's *Physics* and *Categories* and answered a series of questions on the *Physics* as well as writing several works on physics, theology and logic.

FIGURE 6: William of Occam as drawn by Conrad de Vipeth of Magdeburg.

Ripples of alarm flowed out of Oxford, almost as soon as William's work became available. The first indication of trouble comes from the fact that William did not progress along the well-trodden path to becoming a Master of Theology at the university. This was very unusual because, as far as we know, he completed all the requirements. It is not clear who or what prevented William's progression. The principal suspect is John Lutterell, chancellor of the University of Oxford between 1317 and 1322, who composed a tract entitled

*Libellus contra Occam* (*Petition against Occam*). Nevertheless, William continued to lecture and respond to his critics. His *quodlibets* centred on his lectures around 1321–4 after Lutterell had left Oxford. Several other English scholars including a fellow of Merton College called Thomas Bradwardine (1290–1349) accused William of teaching heretical notions. The copying scribes on the Continent had also been busy. As early as 1319–20, William's commentaries had reached France, where they earned the approbation of a French scholar called Francis of Marchia.[4]

To understand why William's ideas were causing such a stir, we have to delve into their source, which goes to the heart of man's relationship with both the world and, if such an entity is accepted, his God.

## The unknowable God

William of Occam's attack on the scholastic philosophy of his predecessors was, in many ways, a continuation of the argument that had been initiated a generation earlier in 1277, when Bishop Tempier of Paris banned debate of doctrines that appeared to restrict God's power to the limits set by Aristotle's logic. Bishop Tempier insisted that the omnipotent Christian God was free to do whatever he liked, irrespective of what Aristotle had to say on the issue.

The bishop's ban did not last but it did stimulate a more critical evaluation of Aristotle's philosophy by scholastics and, particularly, the implication of divine omnipotence. This was an alien concept to classical Greek philosophy as the powers of the Greek gods had always been limited: Poseidon ruled the seas but had little power on land. The Christian God was very different in that He had not only created the universe but had also shaped its rules: he was all-knowing and omnipotent.

The implications of divine omnipotence had been rumbling through the cloisters of Oxford for a generation before William arrived. His predecessor Duns Scotus (1266–1308) had discussed the problem of knowing the difference between right and wrong if God could, arbitrarily, change the rules. William went much further. In an approach that foreshadowed Descartes's dismantling of Western philosophy until he reached the famous dictum of '*cogito ergo sum*', Occam used his razor to strip away everything in medieval philosophy except God's omnipotence.

The problem that William then encountered was that, as well as being all-powerful, an omnipotent God was also unknowable. This becomes clear after we consider that, apart from the law of non-contraction (for example, He cannot simultaneously exist and not exist), an all-powerful God does not need to conform to human reason. For example, he might do unreasonable things, such as make plants on day three of creation (as in the Book of Genesis) before he made light to sustain them on the following day. Although such an order of events might offend Aristotle's reason, it was in God's power to sustain plants, in the dark, for as long as he liked and without providing humanity with any reason for why he chose to do so.

William applied the same kind of reasoning to attack the backbone of philosophy, realism. You will remember that the philosophical realists believed that Platonic Forms, or Aristotelian universals, underpinned the entire world. Cherries were cherries because they shared in the universal of 'cherryness'; fathers were fathers because they were filled with the universal of 'fatherness'.

William wasn't having any of it. He argued that an omnipotent God has no use for universals. If He can make a cherry with the universals of roundness, redness and so on, then He can also make a cherry without these universals. William argued that universals are merely the terms that we use to refer to groups of objects, writing that 'it is vain to do with more what can be done with less . . . Therefore nothing else, beyond the act of knowing, should be posited.'[5] He went on to insist that 'everything that is predictable of many things [universals] is of its nature in the mind', insisting that universals are merely the names that we use to classify objects, hence the term *nominalism* for the philosophical system championed by William in the medieval world.

We can see here our first encounter with William's razor, in the argument that 'it is vain to do with more what can be done with less'. In itself, this was not entirely new. Nearly two millennia earlier, in his *Movement of Animals*, Aristotle had written that 'nature does nothing in vain'. However, rather than an argument for economy in nature, William uses the razor to attack the logic that underpinned universals. He writes that 'The universal is not some real thing having a psychological being (*esse subjectivum*) in the soul or outside of the soul. It has only a logical being (*esse objectivum*) in the soul and is a sort of fiction . . .'[6] Occam insisted that universals have no extra-mental existence so, to avoid muddling up ideas and reality, he urged us to 'not multiply universals needlessly'.[7]

This refusal to 'multiply universals needlessly' is the essential idea behind Occam's razor. We should admit only the minimum number of entities into our explanations or models of reality. Rather than fathers being filled with the essence of fatherhood, Occam insisted that 'We should say instead that a man is a father because he has a son [or daughter].'[8] This statement is so obvious that it may seem trite today and it is hard for us to appreciate its revolutionary impact. Yet, with one stroke of his razor, William dismissed the vast thicket of entities that cluttered the philosophy and science of the medieval world and the world suddenly became immensely simpler and a lot more comprehensible. In contrast, although they acknowledged the virtue of simplicity, Aristotle, Ptolemy, Aquinas and others were content to add complexity whenever it suited them. William was not. This is why, five centuries later, the principle of parsimony was named, not Aristotle's, Ptolemy's or Aquinas's, but Occam's razor.

The dismissal of universals also disarmed that cornerstone of medieval logic, the syllogism. Remember that the logic of 'All men are mortal. Socrates is a man, therefore Socrates is mortal' depends on all men sharing universals such as 'manness' or mortality. Yet, if the only thing in common between Socrates and, say, Plato is a mere word, man, then the fact that Socrates is mortal says nothing about whether Plato or any other man is mortal. The scholastics were horrified. How can they then gain knowledge of the world? For Occam, there was one sure way of discovering whether a man was mortal: fire an arrow into him and observe whether he survives. In Occam's logic, devoid of universals and filled only with individuals, the only way to gain sure knowledge is through experience and observation. This is, of course, the cornerstone of modern science.

It is important however to recognise that this empiricist approach holds no assurance of certainty. A single arrow might prove that Socrates is mortal but it cannot prove that 'all men are mortal'. A hundred arrows felling a hundred men might allow us to propose the hypothesis that all men are mortal but, for Occam, all hypotheses are provisional and probabilistic, vulnerable to disproof by the 101st arrow. For Occam, this was another important distinction between science and religion. For the Franciscan, God's existence was a certainty, but science can only ever consist of hypotheses. Science, he maintained, yields probabilities not proof.

It is not hard to see why Occam's philosophy created a stir. For several centuries, the scholastics had debated the nature of universals and categories, but with a few strokes of his stylus Occam dismissed the entire enterprise as a waste of time; just so much straw, as Aquinas had lamented.

## William dethrones the Queen

Not content with shattering philosophical realism, William went on to attack the scientific pretensions of the proofs of God laid out by Aquinas and others. You will remember that, in four of his Five Ways, Aquinas had argued that Aristotelian causes (material, formal, efficient and final) lead to potentially infinite chains of cause and effect that must be capped with a first cause, God. William pointed out that chains of cause and effect do not necessarily lead to an infinite regress that needs capping. You could, for example, imagine a universe filled with only three objects spending all of eternity continually bumping into each other, generating causes (bumps) and effects (changed trajectories), but remaining just three countable objects. If there is no infinite regress then there is no need for a divine cap. God is an entity beyond necessity, in the Occam's razor sense, so Aquinas's clever argument cannot prove His existence.

As for Aquinas's 'argument from degrees', Occam first admits that a hierarchy of better and greater things must indeed be capped by the very best thing. However, he points out that there is a plurality of 'bests', and each may be capped with its own greatest. For example, Occam and his contemporaries might have argued over which was the most beautiful building, the Notre-Dame de Paris or Canterbury Cathedral; or they might have debated over which was the most poetic cantica in Dante's *Divine Comedy*. Yet, it would have made no sense to argue whether Canterbury Cathedral was better than *The Divine Comedy*. So Aquinas's argument from degrees could have multiple caps such as man, god or donkey depending on which feature is ranked.

However, Occam reserved the main thrust of his arguments against proofs of God's existence for science's principal enemy, *teleology*. You will recall that the telos is the fourth of Aristotle's causes, which in contrast to the others lies in the future not in the past. The telos of pigs was to be eaten. It is anathema to modern science because it undermines its bedrock of causality, which acts from the past to the present into the future. Since we have no access to the future, allowing causes that lie in the future makes science impossible. Nevertheless, Aquinas had argued that God had to be the telos, final cause, of everything in the world. If he was right, then the world would be unknowable as God's teloi are not accessible to mankind.

Occam first admitted that teleological goals may be appropriate for voluntary human actions, such as the construction of a house;[9] but events not caused by intelligent agents have no kind of final cause or telos. He argued instead that 'If I accepted no authority, I would* claim that it cannot be proved either from statements known in themselves or from experience that every effect has a final cause . . . the question "why?" is inappropriate in the case of

---

* Using the conditional was his standard 'get out of jail' clause to insulate his arguments from arguments about divine authority.

natural actions.'[10] Occam insisted that the past or present provides sufficient cause for any event. 'You might ask,' he argues, 'why does the fire heat the wood rather than cool it? I reply that such is its nature.' With this move, William abolished teleology to establish the direction of causality of modern science.* The telos became another entity beyond necessity.

Three hundred years later, and with a lot more pomposity, the great scientists and philosophers of the so-called Enlightenment, or Age of Reason, claimed credit for banishing teleology from science. Yet Occam had put the case against teleology more succinctly and with much less pomposity by insisting that 'No, a natural agent is predetermined by its nature, not by an end.'[11] With teleology eliminated, God was an entity beyond necessity as regards causes in the world and Aquinas's final proof of the existence of God was shattered.

After demolishing philosophical realism – the notion of species and the five most established proofs of God – most scholars would probably have been content with their master's project. Yet William had another incendiary target in his sights: Aquinas's philosophical sleight of hand concerning the central miracle of Christianity, the Eucharist.

## The Queen of Sciences is expelled

You will remember that Aquinas had cleverly manipulated Aristotle's notion of universals in order to incorporate the miracle of the Eucharist into his Christianised science. William of Occam's nominalist philosophy denied that universals existed. With universals dismissed as mere names, William went on to dismiss ten of Aristotle's twelve categories, reducing them to just two: substance and quality. Once again, he used the razor. For example, on quantity, Occam reasoned that an essence of twoness existing in paired objects, such as say two chairs in a room, is illogical. There might be two more chairs in an adjacent room such that, by simply

---

* Some interpretations of quantum mechanics do include retrocausality.

demolishing the partition between the rooms, the chair's twoness would have to transform into fourness. But how could anything real about a chair be affected by the demolition of a wall not connected to the chair? William concluded that 'there is no quantity distinct from substance or quality'.[12] Quantity, in the sense of an Aristotelian category, was an entity beyond necessity that should be eliminated.[13]

However, the category of quantity was where St Thomas Aquinas had hidden bread's taste, smell and texture in the miracle of the Eucharist. There was one bread before the miracle and there remained one Jesus after the miracle. By eliminating the universal of quantity Occam had removed the grounds for Aquinas's incorporation of the miracle into scientia. The Queen of Sciences was toppled from her throne.

## Defending the Third Way

[Of theology] it is not the first or the last or the middle because it is not a proper science . . .

William of Occam[14]

William did not stop with dethroning Aquinas's Queen of Sciences. In another extraordinary step forward, he argued that science and religion are fundamentally and irrevocably incompatible. This followed from his insistence that God transcends human reason so reason is incapable of gaining knowledge of the divine. The only route to God was through faith and the scriptures. Moreover, the scientia roadblock operates in both directions: knowledge of God acquired through faith and the scriptures cannot provide knowledge of the world. Science and theology were thereby two entirely disparate and incompatible modes of human enquiry. Occam wrote that 'it is impossible for the principles [of theology] to be taken merely on faith and the conclusions to be known scientifically . . . it is silly to claim that I have scientific knowledge of the conclusions of theology by reason of the fact that God knows principles that I accept on faith.'[15]

Of course, William was a Franciscan friar. As far as we know, he never doubted the existence of God, nor the central tenets of Christianity. Yet he insisted that his religion came, not from reason, but through faith and study of the scriptures, none of which provides the certainty required of scientia. He thereby embraced *fideism* to argue that 'only faith gives us access to theological truths. The ways of God are not open to reason . . .'[16] Faith was for God. Reason was for science. Although some ancient philosophers, such as the stoics, epicureans and Islamic philosophers,* had argued for a limited separation between science and religion, no one until Occam had made such a clear and cogent argument for the bedrock of modern science: the separation of science from religion. Our largely secular world is the inevitable conclusion of Occam's ruthless logic.

William's razor, coupled with his nominalist and fideist philosophy, essentially opened up a third way between religion and atheism.† It allowed scientists to pursue a secular science while remaining devout. Occam insisted that 'assertions, especially in physics, which do not pertain to theology should not be officially condemned or prohibited by anyone because in such things everyone should be free so that they may freely say what they please.'[17] All the greatest scientists who lived from the fourteenth to at least the nineteenth century were devout Christians who opted to follow Occam's third way.

Nonetheless, all of this was highly unsettling to William's fourteenth-century colleagues in Oxford and London. The theologians who had laboured to build a science out of their discipline

* Al-Biruni (973–1048) contrasted the problems of Indian astronomers having to reconcile their astronomy with the Hindu religion whereas 'the Qur'an does not articulate on this subject [of astronomy] or any other [field of] necessary [knowledge]'.
† Although atheism was far from unthinkable in the medieval world – why would anyone need to prove God's existence if it was never doubted? – and was a view probably held privately by many individuals, any public proclamation of atheism would have, if you did not subsequently recant, earned you a stake on the heretic's pyre.

were outraged, the more astute among them realising that William's insistence on the unknowable nature of God had effectively reduced their discipline to an extended Bible-reading class. The realist philosophers were equally dismayed. They thought that Occam's insistence that universals were mental fictions was the equivalent of telling an economist that money does not exist.

## William gets into trouble

Both Walter Burley (1275–1344) and Walter Chatton (1290–1343), traditionalists who had trained at Merton College in Oxford about the same time as William, delivered lectures and wrote treatises opposing his revolutionary nominalism. One of Occam's students and supporters, Adam Wodeham, recollected taking notes at one of Chatton's classes and rushing them to his mentor (Occam) who scribbled a hasty reply complaining of 'the calumny of some critics'.[18]

In the spring of 1323, when William was about thirty-eight years old, he was called to defend his views at a provincial chapter meeting of the Franciscan order held in Cambridge. He clearly did little to mollify his critics and rumours of his radical ideas continued to leak out of Oxford and London until they eventually caught the attention of the most powerful man in Christendom. The bombshell reached Oxford early in 1324: a summons from the Pope arrived demanding that William attend hearings in Avignon, the papal seat at the time, to answer charges of teaching heresy.

## Avignon

> . . . I am well acquainted with the wickedness of men . . .
> William of Occam, 1335[19]

It is not clear who alerted the Pope to William's potentially heretical ideas but, once again, the former chancellor of the University of

Oxford, John Lutterell, is in the frame. He had travelled to Avignon, probably seeking promotion, in 1323. Pope John XXII asked him to examine Occam's commentaries on Lombard's *The Four Books of Sentences* for potentially heretical opinions. The following day Lutterell presented the Pope with a catalogue of fifty-three 'errors', many concerning Occam's dismissal of Aristotle's category of quantity, the issue that most concerned the miracle of the Eucharist. The Pope responded by summoning William to Avignon to be examined by a panel of six *magistri*, including Lutterell.

The news that one of their number had been accused by the Pope of heresy must surely have caused commotion in the cloisters of London and Oxford. Everybody knew the manner of death that awaited unrepentant heretics. The accused could nearly always avoid the pyre by renouncing their heretical views, but would William? Now in his late thirties, his reputation for obstinacy suggested that he might instead have continued arguing with his accusers while the flames licked around his feet.

William set out for Avignon, soon after the summons. His route would have taken him south to Dover and from there across the Channel on what was almost certainly his first sea voyage. Once in France, William's road would very likely have been via Paris. Did he take the opportunity to meet with scholars and teach there? There is no evidence but, if he had, it might help to account for the enthusiastic adoption of Occamist notions by some Parisian scholars. He would then have taken the old Roman road south to reach Avignon probably by early summer 1324.

The city had become the papal seat after Pope Clement V fled an unruly Rome twenty years earlier.[20] Philip IV of France had offered Avignon as the papal capital and Clement gladly accepted. Yet the city was hardly grand. Lacking sewers, it was famously noxious and colonised by professional thieves, beggars and prostitutes. The poet and humanist Petrarch, who was living in Avignon around the time of William's arrival, described the city as 'Unholy Babylon, thou Hell on Earth, thou stink of iniquity, thou cess-pool of the world . . . Of all the cities that I know its stink is the worst.' Clement was the first of nine popes to reside in smelly Avignon.

It was Clement V's successor, John XXII, who had summoned William. His residence was in the old Bishop's Palace while his new palace, the massive Palais des Papes, with its distinctive Gothic towers was under construction. The old palace was eventually incorporated into the new building so it is likely that William's trial was conducted within the grounds of today's Palais des Papes.

The trial involved a series of hearings in which Occam defended his views in front of the panel of six magisteri, with the Pope probably in occasional attendance. After the hearing, the panel members would deliberate in secret before issuing a report. They first dismissed some of Lutterell's accusations but accepted others and added some of their own. Occam responded by stoutly defending his nominalist and minimalist perspective in a densely argued text, *De Sacramento Altaris*, with the opening argument first questioning 'whether a point is an absolute thing really distinct from quantity' and then concluding that 'a point is not a thing other than a line, or any quantity'. Although the subject matter may seem esoteric, it is extraordinary how rooted William's arguments are in logic, indeed mathematical logic, such as the distinction, if any, between a point and a line. It is not yet science but it is close and it highlights William's efforts to use logic to uproot theology from science's bedrock of empiricism. He goes on to question, perhaps rashly, the competence of the panel, declaring that 'if any of the doctors and saints of the Church had proof that quantity was a thing absolute and distinct from substance and quality, well then it was up to the magisteri to name their sources'.[21]

During this process, William was compelled to remain within the city where he lodged at the local Franciscan convent. It is probably here that he completed his greatest philosophical work, *Summa Logicae* or *Sum of Logic*. It was clearly provocative, as it implied that a single volume, with its strong nominalist stance, contained everything worth knowing in the field of logic. In it, Occam insists that 'logic is the most useful tool of all the arts. Without it no science can be fully known.'

By August 1325, around a year after William arrived in Avignon, it was clear that the trial was not going well. In a response to a letter from King Edward II requesting that Lutterell be returned to England, the Pope replied that the former chancellor was busy in the task of eradicating a 'pestiferous doctrine'. In 1327 Pope John XXII issued a bull charging Occam with having uttered 'many erroneous and heretical opinions'.

However, William's trial, just like his education, was never completed. Instead he became embroiled in another, even deadlier, conflict that had already claimed many lives and, according to several historians, changed the course of European history.

# 4

## How Simple Are Rights?

William of Occam is a giant in the history of thought. He is also one of the most prominent figures in the early development of natural rights theory.

Siegfried Van Duffel, 2010[1]

William was not the only renegade Franciscan residing in Avignon in the 1320s. The order procurator and representative at the papal court, Bonagratia of Bergamo, was incarcerated in the Pope's prison. Its minister general, Michael of Cesena, had also recently arrived in the city and, like William, had been placed under house arrest. Within months, all three men had been excommunicated and were forced to flee. The issue that got them into so much trouble was a fierce debate on whether or not Jesus owned a purse.

As with most apparently trivial medieval debates, the real issue was much deeper than any doubt about the ownership of a purse. It concerned the relationship between the Church, represented by Jesus, and the State, represented by his purse. The origin of the conflict can be traced to the first Christians. Many accepted Jesus's insistence that 'it is easier for a camel to go through the eye of a needle than for a rich man to enter the kingdom of God' and his advice to 'sell all your possessions and give the money to the poor' at face value. They adopted a lifestyle of apostolic poverty that emulated the life of Jesus and his apostles by abandoning money and property to live as itinerant preachers, relying only on begging or charity for shelter and subsistence.

The Roman Church took a very different path. After Emperor

Constantine accepted Christianity as the official religion of the Empire the Christian Church became irrevocably linked with the Roman State. This relationship between Church and State was weakened by the fall of Rome but was re-established in the year 800 when Charlemagne was crowned Holy Roman Emperor by Pope Leo III on Christmas Day in Rome. Thereafter, kings and emperors of western Europe came to Rome to be crowned by the Pope, effectively welding the kingdoms and empires of western Europe, together with its feudal system, to the authority of the Catholic Church.

## Poor, holy and heretical

In the century before William's trial, several renegade Christian groups had sprung up whose members attacked the extravagance of the Church, rejecting the links with the State and embracing the principle of apostolic poverty. These included the Humiliati in Italy, the Waldensians in Germany and the Cathars in the Languedoc region of France.

Most were declared heretical and cruelly suppressed[2] but one of the apostolic poverty groups was, albeit somewhat reluctantly, embraced by the Catholic Church. Giovanni di Bernardone, known as Francis, was born into a wealthy family in Perugia around 1181. After enjoying the usual diversions of moneyed young men, Francis abandoned his inheritance and possessions to live as a wandering beggar and preacher. He and his followers took to wearing a coarse grey woollen tunic tied with a knotted rope, earning them the name of 'grey friars'. Francis and his grey friars travelled around the countryside exhorting anyone who would listen to adopt a life of poverty, penance and brotherly love. He soon built up a substantial and loyal following, prompting Francis to appeal to the Pope to recognise his group as a new itinerant mendicant order. The Pope agreed and Francis's followers became the Order of Friars Minor, better known as the Franciscans. By the time of William of Occam's birth, the order had grown from a small group of eleven to around 20,000 followers.

Yet the Franciscans were not radical enough for some Christians. The founder of another group, the Italian mystic Gerard Segarelli, had been refused entry into the order in 1260 and, in response, carried all his money and possessions into the central square in Parma to distribute every denari, hat, chair and wine bottle among the poor. He grew his beard, wore a white gown and became a kind of white friar walking barefoot from town to town. He soon led a large following of like-minded Christians known as the Apostles and urged whoever would listen to *Penitenziagite!*, roughly translated as 'Repent now!'*

The Church generally had a tolerant attitude to eccentric hermits. Yet, as well as relinquishing his own property, Gerard attacked the Church's wealth. He also insisted that the Church had no privileged access to Heaven, so indulgences (essentially time off in Hell in return for donations to the Church), and the Church taxes known as tithes, were clerical extortion. Unsurprisingly, the Pope declared these views heretical and, in 1300, twenty-four years before William arrived in Avignon, several apostles, including Segarelli, were burned at the stake in Parma.

Yet Segarelli's death only made matters worse for the Church as an even more militant follower, Fra Dolcino, took over the group's leadership.[3] With his partner, Margaret of Trent, whom Dolcino had rescued from a convent, he recruited a large following in northern Italy. The Dulcinites, as they came to be known, were even more radical than the Apostles. They not only rejected the authority of the Church but also refused to accept the authority of the State and insisted that institutions such as property, marriage, laws or serfdom were fictions invented to control people who should be free. Unlike the Franciscans, they welcomed both men and women into their group and lived in cooperative communities rather like a medieval version of hippies.

All of this sounds innocuous today but, by refusing to accept established notions of ownership, feudal dominion and authority,

* The phrase identified the hunchback Salvatore in *The Name of the Rose*, written by Umberto Eco, as a former member of the Dulcinites.

the Dulcinites were effectively setting themselves against the feudal state, as well as the Church. In 1305, Pope Clement V declared a crusade against the group. He promised indulgences to local soldiers, prompting them to attack the rebels, destroying their settlements and pursuing them across northern Italy.

In response, the Dulcinites became more belligerent, raiding villages and monasteries to steal food, money and clothing. In March 1306 they set up a fortified camp on a peak known as Mount Zebello, or Bald Mountain, in Piedmont. The first attack by the crusaders was successfully beaten back, so the bishop resorted to blockading the mountain to starve the besieged group into submission. That strategy worked as the weak and starving Dulcinites were eventually forced to break out of their mountain stronghold, making them easy targets for the indulgence-earning crusaders. In the melee, the soldiers captured Fra Dolcino and Margaret of Trent.

Their trial at Vercelli in Piedmont was swift. Several nobles and gentlemen were so smitten by Margaret's beauty that they offered to marry her if she would only recant. She refused and was sentenced to be burned at the stake, with Fra Dolcino compelled to watch before being tortured and then burned on the pyre himself.

## The holy purse

Although the Franciscans had been founded on the principle of apostolic poverty, by the time Fra Dolcino and Margaret Trent were paraded through the streets of Vercelli, most had abandoned the vagabond lifestyle to live in large well-stocked friaries with kitchens, libraries, dormitories, farms and fishponds. Understandably, many saw this as an abandonment of the founding principles of the order. A solution had been provided in 1279 by Pope Nicholas III who issued a papal bull known as *Exiit qui seminat*, in which he accepted ownership of the Franciscan friaries and all their produce on behalf of the papacy.[4]

Most Franciscans were happy to accept Pope Nicholas's hospitality. However, a more militant faction known as the Fraticelli, infiltrated by escaped Dulcinites, insisted that the bull was a fudge and the friary lifestyle a betrayal of apostolic poverty. The group are today probably best known as the radicals at the edges of Umberto's Eco's medieval crime story *The Name of the Rose*. The apostolic poverty debate plays a key role in the book's plot and its central character, William of Baskerville (played by Sean Connery in the film), was loosely based on William of Occam.[5] Like the Dulcinites the Fraticelli were excommunicated, forcing many to escape to Sicily, at the time ruled by the Holy Roman Emperor Frederick III. Such was the tangled web of allegiances of late medieval Europe that Frederick sent the Christian heretics to Tunis where they enjoyed the protection of its Muslim ruler.

Nevertheless, the Fraticelli hadn't been entirely rooted out. In 1321, while William was lecturing at Oxford, a large group of Fraticelli were arrested in Narbonne and Béziers in southern France for preaching the incompatibility of wealth and holiness. By this time Clement's successor, John XXII, was Pope and he was much less sympathetic to the Franciscans. He ordered the newly elected minister general of the Franciscan order, Michael of Cesena, to interrogate sixty-two of the Fraticelli, asking each whether he considered that Jesus had owned a purse.

When forced to answer John XXII's loaded question, most of the sixty-two rebellious Franciscans backed down and accepted that Jesus owned a purse. They were sent back to their hometowns to publicly renounce their views. Twenty-five who refused were handed over to the inquisitor who persuaded – we are not told how – another twenty-one to recant. The remaining four Fraticelli were burned at the stake, probably witnessed by Michael of Cesena.

However, that was not the end of the matter. On 12 November 1323, John XXII pushed the Franciscans into a tighter corner by issuing the bull *Quum inter nonnullos* declaring the doctrine that Christ and his apostles had no possessions 'erroneous and heretical'. He also revoked Nicholas III's bull *Exiit* that had taken on papal ownership of the friaries. John insisted that, henceforth, Franciscans must accept papal ownership or be considered thieves and trespassers.

## Escape from Avignon

For against the errors of this pseudo-pope I have set my face like the hardest rock . . .

William of Occam, 1329[6]

Late in 1324, just around the time that William arrived in Avignon, his panicked order was gathered in the town of Perugia, not far from St Francis's Assisi, for an emergency conclave to consider their response to the Pope's broadside. The meeting concluded with the drafting of a letter affirming the principle of apostolic poverty. The order's lawyer Bonagratia of Bergamo was charged with taking the letter to Avignon. After arriving in the papal city and delivering the letter he publicly criticised Pope John XXII. William of Occam surely witnessed this event. The Pope responded by exercising his powers of dominion and threw the lawyer into the palace prison. He then summoned Michael of Cesena to Avignon.

The Franciscans appealed to the Holy Roman Emperor,* Louis†
of Bavaria, who was already in dispute with the Pope and was rumoured to be on the verge of crowning his own Pope in Rome, quite possibly even Michael of Cesena. After initially pleading illness, Michael eventually arrived in Avignon in December 1327, where he was publicly admonished by John XXII and placed under house arrest, probably at the same friary where William was being held. This peculiar combination of circumstances, unwittingly engineered by John XXII, brought the cleverest man in Christendom to the attention of the man who most needed his services.

William's deliberations convinced him that the Pope was not

---

* The successor of Charlemagne who had been crowned Emperor in 814 by the Pope. In late medieval times the title was held by a monarch elected by a committee of Prince-Electors. The empire ruled mostly over the German-speaking people but it waxed and waned over the years to encompass other lands, such as Italy.
† Also called Ludwig.

only wrong, but a heretic. With the help of powerful friends, the Franciscans hatched the plan that delivered them to the port of Aigues-Mortes, from where they eventually escaped and where we left them earlier. By fleeing, they had made their situation far more dangerous. Their 'extreme fear' on the deck of Gentile's galley was surely provoked by Michael of Cesena's recollections of the screams of his fellow Franciscans as they burned in Narbonne and Béziers.

## Pisa, Rome, Munich – the Occam tour

Pope John XXII was a famously stubborn man who did not suffer defeat easily. He excommunicated all the fugitives and sent letters to the King of Aragon, the Archbishop of Toledo and the King of Majorca, requesting that the Franciscans should be immediately arrested if they landed in their territory.[7] His focus on westerly destinations had perhaps been seeded in the mind of Lord Arrabley by the wily Gentile during negotiations at Aigues-Mortes. If so, it was a smart ruse because, after an arduous journey of about 250 nautical miles east, lasting about five days, the Franciscans disembarked at the port of Pisa in Italy.

The modern city is now twelve miles inland, but in Occam's time it was a coastal town and a major port for the north Mediterranean maritime trade. From Pisa, the group travelled to Rome where Louis of Bavaria had already crowned himself Holy Roman Emperor and had an unknown Franciscan, Pietro Rainallucci, enthroned as Pope Nicholas V. His Avignon-based rival John XXII responded by announcing a crusade against Louis, declaring his coronation null and void and urging all true Catholics to resist him.

By September 1328 the fickle Romans had become thoroughly sick of 'the Teutons' and Louis was jeered as his retinue departed from Rome to return to Pisa, accompanied by both the rebel Franciscans and Pope Nicholas V. In April the following year the Holy Roman Emperor departed for his seat in Munich, taking with him the Franciscans but not Pope Nicholas. The abandoned pope instead walked all the way to Avignon with a noose around his neck to relinquish his title and beg forgiveness.

William of Occam and Michael of Cesena spent the remainder of their days under the protection of Louis, living mostly in a Franciscan friary in Munich. Occam and his colleagues continued to write articles denouncing John and successive popes. As a fugitive, excommunicate and accused heretic, William's writings in this period shifted from philosophy and science to the conflict that had forced him to flee Avignon and remain an exile for the remainder of his life.

## Simple rights

Although human rights are not a topic usually discussed in books about science, they are, I believe, as necessary for scientific progress as experimental methods or mathematics. Science was clearly possible in slave economies or dictatorships, such as in ancient Greece, or in the feudal societies of late medieval Europe and the Middle East, but it depended on wealth or patronage so was restricted to a privileged few and subject to the whims of wealthy patrons and the demands of the State or the Church. For science to be transformative, it needs a wider base and a kind of scientific democracy in which wealth and power play little or no role in the competition between ideas. This can only happen in societies that provide each individual with the same fundamental rights, including, of course, the right to be wrong.

Which brings us to the nature of rights. What are they? Pope John XXII and William of Occam agreed that ownership works because it provides a right – *ius* or *jus* in Latin (from which we obtain the modern English word, justice) – to use resources such as food or lodgings. But where and how does this right exist? A generation or so earlier, the Augustinian theologian and philosopher Giles of Rome (1247–1316) had staked the Church's case on the account in Genesis of God's ejection of Adam and Eve from the Garden of Eden, after which he conferred on Adam 'dominion over the fish of the sea and over the birds of the heavens and over the livestock and over all the earth and over every creeping thing'. Adam then passed that dominion, effectively of the entire world, on to his favoured descendants, who became the kings, emperors and princes who owned and ruled the entire world. This state of affairs continued until the birth of Jesus who, as both God and living man, took back that ownership and dominion. However, before dying, he bequeathed all rights and ownership to St Peter who passed it on to his papal successors. Subsequent popes then divided their God-given dominion among the Christian monarchs who handed it down to their nobles who shared it with their subjects, though not, of course, their serfs who owned nothing and had no rights. So, far from being empty, the entire medieval world order was held within Jesus's hypothetical purse.[8] In 1493, less than two hundred years after William's conflict with Pope John XXII, Pope Alexander VI divided ownership of the New World between the Spanish and Portuguese crowns, based on this notion of divinely appointed dominion.

Yet the Franciscans had a very different perspective. They insisted that, when Jesus started his ministry, he relinquished ownership of everything to live a life of absolute poverty. So, if he owned a purse, it would have been empty when he handed it over to St Peter. According to their argument, the Church did not even have a rightful claim of ownership over its own churches or lands and certainly none on any wider wealth in the world. But, if the Church's claim to dominion was a fraud, then so too was that of the emperors and princes crowned by the pontiff. The stakes of this conflict were indeed high.

John XXII begins his attack on the Franciscans by reiterating Giles and insisting that 'Dominion over temporal things was not established by primeval natural law, understood as the law common to animals . . . nor by the law of nations, nor by the law of kings of Emperors, but by God who was and is the Lord of all things.'[9] In this John XXII was following the traditional philosophical realist position, which saw natural law as an aspect of the divine reason that saturates the world with God's plan as its final cause. Like the universal of fatherhood, a right was considered to be something that exists independently of the people who claim it. In this sense, it is what is today called an objective right.

In his *Work of Ninety Days* William of Occam reiterated the Franciscan notion of the absolute poverty of Christ. However, if Jesus's purse was indeed empty, then where does ownership or dominion come from? Occam's argument, like John's, begins with theology. He argued that when Adam and Eve were expelled from the Garden of Eden, God provided them, and their descendants, with a natural right to harvest the available resources on earth just as he provided sheep with a right to graze grass. Yet, once again, this simple natural right did not provide ownership. Life was simple. Nobody owned anything but instead existed in a 'state of nature'[10] with basic rights to their own life, sustenance and shelter.

However, after enjoying the ideal 'state of nature' outside the garden, the righteous among Adam and Eve's descendants found themselves having to deal with greedy individuals who consumed more than their fair share. In order to deal with this, they were forced to agree on what was an equitable share of the commonly held resources. From this came the concept of private property, what we call today ownership. Most importantly, this ownership or dominion did not come from God. It was instead an entirely human concept designed to avoid strife. Ownership, according to Occam, was a subjective right, a kind of agreement that existed only in the minds of people who chose to accept it. It had no more objective reality than the notion of fatherhood. It was simply a word or idea.

The communitarian lifestyle was undermined by greedy individuals who stole from their neighbours. To protect themselves against this subjective theft (since ownership is subjective, so is theft), a set of laws were agreed that legislated how private property might be safeguarded, together with suitable punishments for those who broke the laws. To enforce these laws, communities agreed to elect a suitable ruler, perhaps the strongest or wisest among them, who would protect their property with force of arms if necessary. In return, the law-enforcer would receive a larger share of the commonly held resources.

This, Occam proposed, was the origin of earthly dominion or kingship. Essentially, the people loaned an extra share of their natural rights over land or property to their chosen rulers. In later ages, rulers convinced their subjects that this dominion was a real objective right provided by God, the *ius* that underpinned the medieval world order. Yet William insisted that the rights conferred on rulers by their subjects were only loans. Notions, such as kingship or nobility, were mere words. If their ruler misruled then their subjects could reclaim their rights and usurp their rulers. Upending the entire structure of the feudal system, William argued that the authority of rulers derived from the ruled and not vice versa: 'from God through his people'. He also insisted that 'power should not be entrusted to anyone without the consent of all'. William also pointed out that pagans and infidels were, like Christians, descended from Adam and Eve. They therefore inherited the same natural rights as Christians and, like them, had the right to invent their own laws and elect their own legitimate rulers.[11]

Returning to the Franciscan dilemma, William insisted that despite relinquishing the man-made concept of ownership, they nevertheless possessed their natural God-given right to use resources when in need. He insisted that these natural rights could not be annulled by pope or emperor, even voluntarily, as 'no one can remove . . . the rights and liberties conceded to the faithful by God and nature'; and 'no one can renounce the natural right of using'.[12] Although many lawyers and philosophers have contributed to the notion of subjective rights, the twentieth-century French historian of law Michel Villey had no doubt who was responsible. He wrote that the 'Copernican moment' in the history of law was associated with 'the whole philosophy professed by Occam . . . that is the mother of subjective rights'.[13]

William's *Work of Ninety Days* was extensively copied and, two hundred years later, influenced many of the key figures of the Reformation. The English king, Henry VIII, had his own copy in the library at the Palace of Westminster and consulted it, even annotating its pages, to help build his case for divorcing Catherine of Aragon. During the English Civil War the copy made its way to Lanhydrock House in Cornwall, now owned by the National Trust, where it resides to this day. Occam's nominalist notion of subjective rights also went on to influence key figures of the political Enlightenment, such as Hugo Grotius the Dutch humanist, poet, playwright and lawyer,[14] and through them Thomas Hobbes, George Berkeley and the nineteenth-century materialists who, like Occam, insisted that the right to property or to rule is a human invention. As Karl Marx noted, 'Nominalism was one of the principal elements of English materialism, and in general is the first expression of materialism.'[15]

# 5

# The Kindling

We will return to Oxford where William's ideas are providing the kindling to ignite a brief though brilliant blaze of science in the college cloisters. Where he studied in Oxford remains a mystery but one of its oldest colleges, Merton, established for students of theology about fifty years earlier, is a likely candidate. After William's hasty departure from Oxford, and despite his heretical status, his ideas continued to be studied at Merton. For example, in 1347, a fellow of Merton, Master Simon Lambourne, left a collection of Occam's essays to the college including one of his commentaries on Lombard's *The Four Books of Sentences*.[1] Most remarkably, in the decades immediately following William's departure from Oxford, a group of scholars known as the Merton Calculators achieved widespread fame, not in theology, but in their innovative application of mathematics to the natural sciences, a move probably inspired by William.

None of the Calculators refers directly to William or his work – he was an accused heretic and excommunicate at this time – but it is not hard to discern his influence, particularly his enthusiasm for a particular mathematical heresy.

## *Squaring the circle*

You will remember that Aristotle was keen on categorising. He divided his universals among ten categories that included substance, quantity, quality, time, place, passion, action and so on. He further complicated matters by prohibiting lines of reasoning or proofs from one category being applied to another. For example, the category of quantity included numbers but no substances; whereas the quality category was used to characterise objects of substance including their habits, such as whether they tended to fall (stone), to rise (smoke) or to melt (ice). Aristotle argued that different rules operated in different categories and, in particular, mathematics could only be applied to objects without substance, such as circles, triangles or the heavenly bodies. He wrote that 'arithmetic and geometry are not concerned with any substances'.[2] Numbers or geometry were therefore unsuitable tools to describe, say, the hotness of an object or the trajectory of an arrow. Instead, only qualitative terms, such as warm, cold, curved or straight could be applied.

Mathematics is, of course, fundamental to modern science. Physics would be unthinkable without it but it is also a vital tool in chemistry, biology, geology or meteorology. All of these were lumped under the term 'natural science' in the medieval world and were out of bounds to mathematics because they all concerned 'substances'. This was a severe impediment to scientific progress, particularly as mathematics is the doorway to simplicity. How do you measure the third side of a right-angled triangle? You don't need to if you know the length of the other two sides and Pythagoras's theorem. This is what mathematics delivers to science: a simpler and thereby more comprehensible and predictable world. For Aristotle, that tool was only available to insubstantial objects such as light, the universals of triangles or heavenly bodies.

The Greek philosopher did however allow a limited amount of what he called *metabasis* in which proofs of one science could be co-opted to a *subalternate* science that was considered to be subordinate to the higher science. For example, the music played on stringed instruments was considered to be subalternate to mathematics as harmonies are predictable from the ratios between the length of strings and the notes they play when plucked. If a string is plucked to give a particular note then a string of half the length will produce a note that sounds one octave higher. So the octave musical interval has a mathematical proportion of 2 to 1; whereas string lengths in a ratio of 2 to 3 are a perfect fifth apart. However, apart from these few exceptions, Aristotle imposed a general ban on metabasis in the sciences.

A related Aristotelian restriction was his insistence that different mathematical objects were incommensurate. For example, a circle cannot be compared with a square since, he claimed, it is impossible to use either numeric or geometric methods to determine the area of a square that would equal that of a circle. Such an attempt at *squaring the circle* would, he insisted, be an offence under the metabasis-prohibition rule. Similarly, every geometric object was considered to be underpinned by its own existing universal and thereby no more comparable than the taste of cheese could be compared with the sound of a lute.

Categories, metabasis and incommensurability survived the fall of the ancient world to be transmitted, often via Arab scholars, to Western scholastics. So when Islamic or European medieval philosophers considered a subject such as motion, their first question was 'In which category of being does motion belong?' It was vital that they answered this question because its answer determined the nature of the science they could apply. Unfortunately, Aristotle's categories were so numerous and woolly that the scholastics hardly ever got beyond this question. Aquinas's mentor Albert the Great wrote an extensive discussion on the question of the category of motion in his commentary on Book 3 of Aristotle's *Physics*, citing both Aristotle and the opinions of the Arab commentators.[3] He deliberated over whether motion might fall into the category of action, passion, quantity, quality, place or whether it represented an entirely new category all on its own. Unsurprisingly, neither he nor any of the scholastics ever managed to reach a conclusion.

Occam's dismissal of eight of Aristotle's ten categories as entities beyond necessity delivered the immediate benefit of banishing most of the metabasis prohibitions. As for mathematics, William used his nominalist razor to attack the notion that mathematics is based on Platonic Forms or universals of triangles, circles and numbers existing in some perfect realm. He wrote that 'if [mathematical] relations were real then when I moved my finger and its position is changed in relation to all the parts of the universe' [then] 'heaven and earth would be at once filled with accidents'.[4]

Since numbers, shapes or geometrical objects were mere mental tools, he argued, there should be no restriction on their application. For example, in the prologue to his *Ordinatio* completed just before his departure for Avignon in 1324, Occam discusses the relationship between the sciences and mathematics and argues that many of those sciences considered out of bounds for mathematics by Aristotle, such as medicine, nevertheless find good use for mathematical concepts. For example, a physician might provide a different prognosis for a wound depending on whether it was a straight cut from a sword (good prognosis) or a round piercing from a spear (poor prognosis).

Similarly, Occam blew away the ban on comparing supposedly incommensurate quantities such as straight and circular lines. He simply pointed out that a rope coiled into a circle could be uncoiled to determine whether its length is longer, shorter or equal to a straight rope.[5] Ignoring centuries of philosophical deliberations, William broke through to the modern perspective of a nominalist science based on experience.

## The Calculators

Thomas Bradwardine, William's contemporary, was first to allow his stylus to take advantage of Occam's relaxation of Aristotle's prohibitions to the study of motion. For Aristotle, motion was just a form of change alongside growth or decay. He pointed out that motion was only possible when the force acting on a body exceeds the resistance to motion but never attempted to encode his principle in mathematical form. Ignoring the metabasic prohibition in his *Tractatus de Proportionibus* written around 1328, Bradwardine imported the concept of mathematical proportions in musical intervals to argue, correctly, that it is the mathematical ratio between force and resistance – a number – that determines the quantity of motion.[6] This was a revolutionary move as, for the first time, it applied mathematical reasoning to objects known to be made of matter.

Bradwardine went on to become an influential diplomat and Archbishop of Canterbury yet, back in Oxford, his mathematical baby steps were followed by another generation of Merton scholars, including John Dumbleton (c.1310–c.49), William Heytesbury (c.1313–c.72) and Richard Swineshead (who died around 1364). All overlapped at Merton in the period between 1330 and 1350, so it is easy to imagine the group poring over manuscripts together by candle-light in the stone-cold college library.* Heytesbury and Dumbleton were heavily influenced by William of Occam's nominalist logic.[7] Yet once again, Occam's biggest impact on science at this time was his freeing of mathematics from its philosophical shackles.

Heytesbury, who would later be known simply as 'The Calculator', wrote *Regulae solvendi sophismata* or *Rules for Solving Logical Puzzles* in 1335 in which he even invented a kind of semi-mathematical metalanguage that he applied to lots of problems forbidden under the metabasis restrictions, such as the relationship between weight and resistance to motion.[8] In typical scholastic style he posed questions, such as whether 'there is either a maximum weight that Socrates can lift at velocity A in medium B or a minimum he cannot lift'.[9] The most important advance he and his fellow Merton Calculators made was their definition of speed as a relationship between distance and time. Aristotle had never attempted any mathematical definition since he considered motion to be a complex notion involving change in place, time, location and position, all separate categories of being and thereby incommensurate. The Merton Calculators metaphorically uncoiled Occam's rope to define speed simply by dividing the distance that an object moves by the time it takes. The definition is often credited to Galileo,[10] but the Merton Calculators came up with it three centuries earlier.

## Making laws with Occam's razor

With a mathematical description of speed under their belts, Heytesbury and his colleagues went on to discover the first law of

---

\* Fires were usually not allowed in libraries because of the incendiary nature of books.

modern science, the mean-speed theorem. The theorem states that the distance travelled by an object uniformly accelerating from rest is equal to the distance that the object would have travelled if it had been travelling at its average (mid-point or mean) speed for the same time. So if a donkey smoothly accelerated from rest to a trot of ten miles an hour, in one hour, then the distance it would have travelled would be the same as if it had plodded along at a steady five miles an hour, for an hour, that is, five miles.

Scientific and mathematical laws are crucial to our story because beneath their steely exterior, they are the purest expressions of Occam's razor. Remember Einstein's insistence, mentioned in the Introduction, that 'The grand aim of all science [is] to cover the greatest number of empirical facts by logical deduction from the smallest possible number of hypotheses or axioms.'[11] Scientific laws, for example about light, motion or heat, are all ways of covering 'the greatest number of empirical facts' from simple hypotheses and axioms. Their value can be grasped if you imagine how Aristotle might have responded if you had asked how far a donkey would have travelled if it had smoothly accelerated from rest to ten miles per hour in one hour. He would probably have told you that it all depends on the donkey's material, its formal, efficient and final causes of motion and the particular categories in which those causes were categorised. The donkey would probably have expired before Aristotle had finished his answer.

Yet, if asked, Heytesbury and his colleagues would have replied it was simply half the donkey's final speed divided by the time it had taken to reach that speed. Moreover, if you amended your question to ask about the acceleration of a goat, cow, comet, scholar or arrow, objects made from quite different substances and belonging to different categories of being, then they would have told you that it made not the slightest difference. When calculating the answer details such as material were entities beyond necessity.

The mean-speed theorem is remarkably useful. However, it does have an important limitation. The Merton Calculators only described motion; they did not attempt to account for motion by providing its cause. In today's terminology, we would call the mean-speed theorem a kinematic theory of motion. There is nothing intrinsically wrong with a kinematic theory; they remain useful today. However, they have nothing to say about the future or the past unless either is exactly like the present. For science to be able to predict an uncertain future, it also needs to be able to deal with change and, for that, it needs to develop models that incorporate causes. The next advance in the study of motion was taken by Occamist scholars in the city where William probably enjoyed a stopover on his way to Avignon.

## What causes a cause?

Jean Buridan was born to a humble family in the diocese of Arras in Picardy, France, sometime around 1300. The clever young child came to the attention of a wealthy benefactor who paid for his schooling at the Collège Lemoine in Paris and then at the University of Paris. In about 1320 he obtained a licence to teach and quickly advanced through the academic system. He was so successful that he was soon described by colleagues as a 'celebrated philosopher' and was twice appointed rector of the University of Paris. He would certainly have been there when William of Occam may have passed through its cloisters.

Sadly, we know very few facts about Buridan's life, except for several scandalous rumours. Most centred on his reputation as a philanderer. In one story he is said to have struck the future Pope, Clement VI, over the head with a shoe while both were competing for the affections of a German shoemaker's wife. Another tale tells of King Philip V of France having Buridan tied into a sack and thrown into the Seine after discovering that the philosopher had been having an affair with his wife. One of his students, apparently, saved him from drowning.

Most of these stories are probably apocryphal but we do know for sure that Buridan was one of the greatest scholars of his age. He wrote commentaries on Aristotle's works, including *The Organon, Physics, On the Heavens, On Generation and Corruption, On the Soul,* and *Metaphysics.* Buridan's major work was *Summulae de dialectica,* which became the standard textbook that spread William of Occam's nominalist logic across European universities where it became known as the *via moderna* or 'new way'. In the words of the historian T. K. Scott: 'What Occam had begun, Buridan continued . . . If Occam initiated a new way of doing philosophy, Buridan is already a man of the new way. If Occam was the evangel of a new creed, Buridan is inescapably its stolid practitioner . . .'[12] The *via moderna* opposed the conservative, entity-bloated scholastic tradition of the *via antiqua* (old way) of philosophers such as Aquinas or John Scotus, striving instead for a simpler, less cluttered philosophy that was largely based on Occam's nominalism, his separation of science from theology and ruthless application of his razor.

Buridan's most influential scientific breakthrough was his discovery of a revolutionary way to describe the causes of terrestrial motion, such as the flight of an arrow. Aristotle described these motions as violent and required them to have a preceding material, formal and efficient cause. Yet, even with so many causes, Aristotle's system failed to account for why an arrow continues to fly through the air long after it has left the bow. Perplexed, Aristotle responded in his usual way by throwing in more complexity. He proposed that, after receiving its initial thrust from the bowstring, the moving arrow generated a kind of whirlwind in the air surrounding the arrow, which continued to propel the arrow along its path.

William of Occam had already spotted the flaw around a decade or so before Buridan's deliberations.[13] He pointed out that two arrows travelling in opposing directions could fly right past each other in mid air. At the near-miss point, Aristotle's whirlwind would need to be pushing in two contrary directions, which does not make sense. Jean Buridan instead proposed that the moving bowstring imparts a quantity of *impetus* to the arrow. This impetus remains attached to the arrow, as a kind of fuel that pushes against air resistance, until it is exhausted and the arrow reverts to its natural motion of falling to the earth.

The concept of impetus was not entirely new. It had been introduced in the sixth century by the Byzantine philosopher John

Philoponus (c.490–c.570) and further elaborated by the Persian scholar Ibn Sina (Avicenna) born in 980 CE. However, what made Buridan's concept truly revolutionary was its mathematical definition. Buridan proposed that an object's impetus could be calculated by multiplying its weight times its speed. This is similar, but not identical, to the modern concept of momentum.*

Buridan's was the first causal law of motion to be described mathematically, making it the progenitor, whether direct or not, of most of the scientific laws that shape the modern world. Like the Merton Calculators, Buridan was attempting to 'cover the greatest number of empirical facts by logical deduction from the smallest possible number of hypotheses or axioms'.

Before moving on, I want to explore one final question regarding the nature of impetus. Would Buridan have understood the motion of an arrow any less if he had proposed that, instead of impetus, the bowman's bow delivered an angel to the arrow and, thereafter, the arrow's motion was powered by the beating of angel wings until the angelic driver was exhausted? The question appears ludicrous to us but it certainly was not for the medievals. To most of them, an angel was far more real and present in the world than impetus.

For the time being, we will leave this question hanging. It is however one to which we will be returning as, when generalised, it is central to the role of Occam's razor in science.

## The earth moves (maybe)

In his surviving *Ordinatio*, William of Occam noted that, to an observer standing on the deck of a ship travelling along a tree-lined shore, 'The trees . . . appear to move'. He went on to argue that these propositions are equivalent: 'the trees . . . are seen successively in different distances and aspects by an eye moved with the ship's motion', and 'the trees seem to the eye to move'.[14] Occam was

---

* Momentum equals velocity (a vector that includes direction) times mass.

pointing out the relative equivalence of motion and rest: it depends on your perspective. Occam used this observation to argue that motion, like a universal, is not an existing thing but a relationship between objects. Buridan realised that this relativity of perception could also have heavenly implications.

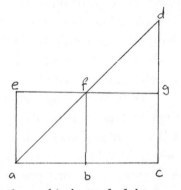

FIGURE 7: Buridan's graphical proof of the mean-speed theorem.
The horizontal axis *ac* is 'time' whereas the vertical axis *cd* is uniformly
'accelerating speed' so the total distance travelled is given by the area
of the triangle *adc*. Buridan pointed out that if the point *f* is equidistant
between *a* and *d* then the triangle *fdg* is the same size as the triangle
*aef* so the rectangle *aegc* has the same area as the triangle *adc*.
But its area is equal to the distance travelled if travelling at
an average speed for the entire journey.

In Buridan's motion theory, the bowman's bow imparts a quantity of impetus that causes the arrow to fly through the air. Yet arrows eventually fall. Buridan reasoned that this was because the bow imparts only a limited quantity of impetus that acts against air resistance until exhausted. However, he went on to speculate that 'Impetus would last for ever if it were not diminished and corrupted by an opposing resistance or a tendency to contrary motion.'[15] This is actually very close to the modern concept of inertia, usually credited to Galileo. Moreover, Buridan also proposed that 'In the celestial motions there is no opposing resistance',[16] so after God gave them their first injection of impetus, the heavenly bodies could keep moving for ever. This was already a big step towards a mechanical heavens operating according to terrestrial laws (as William had proposed). However, Buridan pondered an even more revolutionary, and potentially heretical, idea by importing William of Occam's observational equivalence of motion and rest to argue that the earth, rather than the stars, might move.

Like everyone else, Buridan had noted that the stars appear to rotate around the earth every day but he also realised that this could be a matter of perspective. If the earth was turning, then their daily orbits would vanish. He wrote that:

> Just as it is better to save the appearances through fewer causes than through many . . . Now it is easier to move a small thing than a large one. Hence it is better to say that the earth (which is very small) is moved most rapidly and the highest sphere is at rest, than to say the opposite.[17]

Buridan stilled the motion of thousands of stars by allowing a single object, the earth, to spin: Occam's razor. However, the French scholar was also clever enough to spot a problem. If the earth really was rotating from west to east at great speed then an arrow shot vertically into the air should fall east of its starting position. Since that did not happen, Buridan concluded that the earth must be still and the heavens do indeed turn.

This was sound reasoning, but of course wrong. Its correct solution was provided by one of Buridan's students, fellow Occamist and follower of the *via moderna*, Nicole Oresme (c.1323–82). Oresme enjoyed an even more illustrious career than his master, becoming tutor to the future Charles V of France (1338–80), and later appointed Bishop of Lisieux. While studying with Buridan in Paris, Oresme studied the work of the Merton Calculators and, once again ignoring Aristotle's metabasis prohibition, used geometry to provide a graphical proof of their mean-speed theorem (Figure 7). He followed his master in using Occam's razor to eliminate the daily rotation of the stars by claiming that 'something done by several or large-scale operations which can be done by fewer or smaller operations is done for nought'. However, unlike his teacher,

Oresme solved the problematic arrow conundrum by pointing out that an arrow fired vertically from the deck of a moving ship nevertheless lands on the deck. This is because, he reasoned, the fired arrow shares in the ship's horizontal impetus and so will continue moving along with the ship even after the arrow had left the bow. Oresme argued that an archer on the surface of a moving earth is in the same situation as the sailor on the deck of a ship and 'For this reason the arrow returns to the place on the Earth from which it left.'

However, like his master, Oresme remained unwilling to take the great leap into a simpler cosmos. He argued that, since reason alone could not distinguish between the two perspectives of a rotating earth or rotating celestial sphere, he abandoned the razor for the scripture. In the book of Joshua in the Bible Oresme found a passage in which God commanded the sun to stand still in the heavens to provide Joshua with more daylight hours in which to slaughter his enemies.

Despite Oresme's theological foot-dragging, William of Occam's *via moderna* had, by the 1340s, taken huge strides towards escaping the tangled thickets of Aquinas's scientific theology. If progress had continued then the Industrial Revolution might have happened in the sixteenth rather than the eighteenth century. Sadly, a microbe brought the last scholastics down to earth.

## The pestilent years

In the year 1347, the Mongol horde was laying siege to the port of Kaffa on the Crimean peninsula demanding the surrender of Genoan merchants accused of murdering the city's overlord. When the besieging army fell victim to a mysterious but deadly illness the Genoans thanked God. Their thanks were short-lived. The Mongols catapulted corpses into the city. Its citizens fell victim to the malady prompting the Genoans to sail back to Italy. Their ship stopped at probably the most populous city in the world at that time, Constantinople. Within weeks, thousands of its inhabitants were dead. The ship's next stop was Messina, Sicily, in October 1347, by which time most of the crew were dead. Twelve surviving, though sick, Genoans were prevented from disembarking but carried on the backs of its rats the disease jumped ship anyway. Within months, all the major ports of Europe were affected. Within a handful of years, more than half of all Europeans were dead, including Thomas Bradwardine, Jean Buridan and William of Occam. Although the universities mostly survived, a shortage of teachers caused a collapse of basic education, and literacy levels plummeted.

The first epidemic burned itself out within four or five years but, over the following decades, outbreaks of the plague continued to devastate Europe with grim regularity. Looking for someone to blame, frightened rulers and citizens targeted the Jews and thousands were murdered. Many others believed that man's wickedness was the cause of the divinely inflicted suffering so, in a bid to appease an angry God, donned sackcloth and ashes to wander from town to town to gather and mutually scourge each other with iron-tipped whips. Yet flagellating, penance, prayer or purges did nothing to placate the fierce God. None were spared. Medieval Europe plunged from the rural idylls portrayed in *Très Riches Heures du Duc de Berry* (*The Very Rich Hours of the Duke of Berry*) into the hellish visions of Hieronymus Bosch. With death around every corner, the scholastics abandoned scientific speculation and took to their prayers. It would be more than a hundred and fifty years before anyone else in medieval Europe took any serious interest in science.

# 6

## The Interregnum

The year is 1504 in the city of Florence where the Tuscan artist Leonardo di ser Piero da Vinci, known today as Leonardo da Vinci (1452–1519), is packing his books. Since the great plague that swept across Europe killing so many of its inhabitants it has been 157 years. The Black Death had been particularly severe in Florence, wiping out three quarters of its population between 1347 and 1348. Yet, by the sixteenth century, outbreaks are rare and less severe.[1] The city is recovering, indeed thriving, as one of the fastest-growing cities in Europe.

Leonardo had been born out of wedlock to the notary Piero da Vinci and his domestic servant Catarina. They lived in the foothills of Montalbano outside the town of Vinci. In the mid-1460s, his family moved to Florence where the young Leonardo was apprenticed to the studio of the sculptor, goldsmith and painter Andrea del Verrocchio. It wasn't long before Leonardo's remarkable talent caught the attention of wealthy and influential patrons who commissioned projects, such as the unfinished *The Adoration of the Magi* in the monastery of San Donato in Scopeto in Florence, now in the Uffizi gallery. He moved to Milan in 1482 where he painted the *Virgin of the Rocks* for the Confraternity of the Immaculate Conception and his extraordinary *The Last Supper* for the monastery of Santa Maria delle Grazie in Milan.

Leonardo's commissions continued in the succeeding decades, not only for works of art but also architectural and engineering projects. In 1499, he devised a structure of moveable barricades to protect the city of Venice from flooding; and, three years later, worked with Niccolò Machiavelli to design a system for diverting the River Arno. That project was an expensive disaster that resulted in the loss of eighty lives. Undeterred, the Florentine Signoria (local government) commissioned Leonardo to work, with Michelangelo, on the painting of its Palazzo Vecchio. Later that year, however, Leonardo's father died and he made plans to travel to Vinci. Before his departure, he packed up all his books and manuscripts and prepared two catalogues. The first was entitled 'A record of the books I am leaving in the locked chest'; the second catalogue listed the books kept 'In the chest at the monastery', presumed to refer to Santa Maria Novella.[2] These library catalogues were stored along with the books they described.

Leonardo is, of course, most famous for his paintings, considered among the greatest masterpieces of Western art, but he was also a true Renaissance man who made thousands of pages of notes that included wonderfully naturalistic drawings of rock formations, crystals, birds, fossils, animals, plants, human anatomy and both real and imaginary machines. He carefully stored these notes and, after his death in 1519, they were bound together into several notebooks, known today as the *Leonardo Codices*. Several have since been lost but many have survived and are in private collections or museums.

Over the years, the *Leonardo Codices* have mostly been admired for their artistry but in the nineteenth century historians of science began to take an interest. The notes were doubly difficult to decipher as he wrote in Latin in a cursive shorthand mirror writing that is almost indecipherable. One of the documents, Codice A, had been held in the Biblioteca Ambrosiana in Milan until 1796 when it was purloined by Napoleon during his invasion of Italy and brought to the Bibliothèque de l'Institut de France in Paris where it remains today. There, in the early twentieth century, a French physicist and historian of science, Pierre Duhem (1861–1916), was laboriously picking though Leonardo's difficult text when he was astonished to recognise familiar mathematical laws concerning motion and falling weights, together with ideas related to the conservation of energy.[3] Another document included a drawing of a bird's wing with annotations that read: 'The hand of the wing is that which causes the impetus, and then its elbow puts itself with the edge forward not to hinder the motion caused by the impetus.'[4] He was all the more surprised because the prevailing dogma of the early twentieth century was that science had virtually disappeared during the 'Dark Ages' after the fall of the Roman Empire and had not re-emerged until the so-called Age of Enlightenment in the seventeenth century. Leonardo's notes had been written in the fifteenth century: where had Leonardo's understanding of sophisticated scientific principles come from?

Duhem guessed that the answer might lie in the contents of the library chest, but sadly they, and the books within them, had long since vanished. The library catalogues had, however, survived. One is held in Madrid.[5] Duhem was able to access a copy where he found a list of book titles on widely diverse science subjects from medicine to natural history, mathematics, geometry, geography, astronomy and philosophy. Many were familiar works by ancient Greek philosophers including Aristotle, Ptolemy and Euclid, but the list also included less well-known works by medieval scholars, such as *De Caelo et Mundo* (*Of the Heavens and the World*) by Albert the Great. Duhem located any surviving copies he could and in them was able to discover more references to the scientific notions in Leonardo's notes, such as impetus. Many of the original texts were commentaries on earlier works by the Parisian Occamist scholars Jean Buridan and Nicole Oresme. Further detective work by Duhem and, later, by Ernest Moody (1903–75) followed the tracks of Leonardo's scholarship across the Channel to the group of medieval English scholars known simply as 'The Calculators' as well as to a movement known as the *via moderna* that had been inspired by William of Occam.[6] Rather like the rediscovery of Greek texts in the twelfth and thirteenth centuries, Duhem and his colleagues uncovered an entirely forgotten period of science and he concluded that: 'In the mechanical work of Leonardo, there is no essential idea that does not come from the geometers of the Middle Ages.'[7] The plague that had wiped out the practitioners of the *via moderna* clearly did not destroy their ideas.

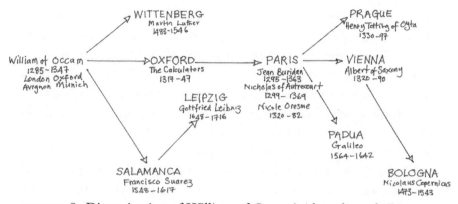

FIGURE 8: Dissemination of William of Occam's ideas through Europe.

There is no reason to believe that Leonardo had any special access to the ideas and philosophy of the *via moderna* in the fifteenth century so it is likely that thousands of other scholars were familiar with Occam's razor and the science that it inspired. Yet it remains a puzzle how the advances of the *via moderna* had been transmitted through the centuries before the invention of the printing press. Subsequent research has uncovered two main routes, one for each of the two great cultural revolutions of the late Middle Ages.

## The southern route to the Renaissance

Seventy-two years before Leonardo was born, on a night sometime around 1380, one of Florence's greatest musicians and composers, Francesco Landini (c.1325–97), was having a dream. In it, he was visited by a famous English friar. Landini was the most famous and innovative musician in Florence, indeed in the whole of Italy. The son of a painter, Jacopo del Casentino (c.1310–49) of the school of Giotto, Francesco would probably have followed his father into the artist's studio if, as a child, he hadn't been struck blind by a bout of smallpox. The young man turned his creative talents to music, poetry and musical-instrument making. His singing voice was legendary. In *Il Paradiso degli Alberti*, the writer, mathematician and humanist philosopher Giovanni da Prato said Landini's playing was such 'that no one had ever heard such beautiful harmonies, and their hearts almost burst from their bosoms'. He mastered a wide range of instruments, from the medieval rebec and the flute to the hand-held organ called the organetto. His skills had also been enlisted to construct musical instruments, including organs for both the Basilica della Santissima Annunziata and the Cathedral in Florence. He even invented his own instruments, including a variant of the lute called the syrena syrenarum.

Landini's greatest fame was as a composer of madrigals. Most were for two voices and blended French and Italian influences to create a novel style that was highly sought after at the fashionable gatherings of the Florentine cultural elite. There, the wealthy, talented, beautiful, powerful or clever met to recite poetry, discuss the latest art works, or listen to music often composed, and sometimes played and sung, by Landini. Whenever the citizens of Florence invited the great musician and composer, they also knew to expect some philosophical interludes between the verses. Philosophy was Landini's other great love, particularly the revolutionary nominalism of William of Occam whose ideas had trickled into Italy along the well-trodden trade, pilgrim and diplomatic routes. Landini even incorporated Occam's philosophy into his lyrics. In his song 'Contemplar le gran' (Contemplate great things) he wrote that 'The articles of Christian faith . . . should be accepted as such. They cannot be proved by reason; nor can they be made the basis of knowledge. Science and theology are essentially different and must not be confused.' In another verse, he sang that 'It is good to meditate upon the great works of God, but it is unnecessary to explain them.'*

The *trecento*, the century that preceded the pinnacle of the Renaissance's *quattrocento*, was a time of intellectual upheaval in Italy as its culture edged away from the old medieval world towards an uncertain future. In a letter Landini wrote in Latin verse to a friend in Avignon around 1380, the composer describes how the ghost of William of Occam came to visit him in a dream to complain

---

* Recordings of Landini's music continue to be available today.

of the 'savage dogs' who attacked the rational philosophy of the so-called 'northern barbarians' and how such a person 'hates logicians like he hates death'.[8] The 'savage dogs' were actually some of the fashionable thinkers of the Italian Renaissance who were already turning their backs on the scholastic philosophers. Despite being a member of this movement, Landini rushed to William of Occam's defence. At the end of a long invective against 'an ignorant man' who has been rousing the 'ignorant masses' against the great philosophers of the past, the ghost of Occam is alerted to the approach of the day by the sounds of street vendors and the 'venerable shadow disappeared into the air'.

This wonderful story, uncovered only in 1983, demonstrates how, by 1380, William of Occam's philosophy had seeped out of Oxford, Avignon, Paris and Munich, to infiltrate the fast-beating heart of the trecento. How and why Landini became familiar with William of Occam is not clear but there are a number of possible routes. The Tuscan poet Petrarch lived in Avignon while William of Occam was confronting the Pope in the city, and later travelled to Florence and was probably known to Landini. Despite his blindness, Landini also travelled widely and he may have come across Occam's teachings through an influential textbook on nominalist logic entitled the *Perutilis logica* (*Very Useful Logic*) by one of Buridan's Occamist students, Albert of Saxony, which had been widely copied and circulated among the major centres of learning in Europe including Prague, Paris, Oxford, Vienna, Bologna, Padua and Venice.

As we have already discovered, the scholastic copyists provided a surprisingly rapid route for transmission of manuscripts across Europe. Yet manuscripts were extremely expensive, making them a luxury to which access was restricted to the clergy or the wealthy literate elite. That all changed in 1445, around a hundred years after Occam's death and sixty years after Landini's dream, when Johannes Gutenberg invented the modern printing press. First off the press was the famous Gutenberg Bible, printed in Mainz in 1455. In the following decades, print shops sprang up all over Europe. Prior to the printing press, only about 30,000 books existed in the whole of Europe; by 1500 there were more than 9 million in circulation. As they became more numerous, books became cheaper and accessible to wealthy merchants, scholars and craftsmen. Literacy rocketed, increasing the demand for new books, and as the Bible-printing market became saturated, printing houses scrambled to find vellum manuscripts whose text could be copied into typescript.

Next off the presses came printed copies of theological works by Augustine or Aquinas and others, and the works of ancient philosophers including Aristotle, Galen, Ptolemy and Euclid. One of Leonardo's chests contained a copy of Euclid's *Elements* printed in Venice by Erhardus Ratdolt in 1482. In 1471, the first printing house specialising in scientific texts was set up in Nuremberg by Regiomontanus (1436–76), a key figure of the German Renaissance. He printed the astronomical textbook on Ptolemy's astronomy *Theoricae Novae Planetarum*, based on lectures delivered by his teacher, Georg von Peuerbach. Venice was also an important publishing centre and Leonardo's library included a copy of Sacrobosco's *Tractatus de Sphaera* published in the city in 1499.

With theology and ancient philosophy texts soon exhausted, printers turned their attention to modern works on the science and philosophy of the *via moderna*. Most of William of Occam's major philosophical and theological works were printed around this time. His *Ordinatio*, on the first book of Peter Lombard's *The Four Books of Sentences*, was edited for the first time in Strasbourg, in 1483, and then re-edited and printed in Lyons, in 1495, together with his *Reportatio* on Lombard's *Sentences* II–IV. *Quodlibeta Septem*, the record of his disputations, together with his *De Sacramento Altaris*, was printed in Strasbourg in 1491, while his *Summa Logicae* was printed in Paris, in 1488.[9] Many of his most influential works, including his *Summa Logicae*, were also printed in Bologna between 1496 and 1523 by Benedetto Faelli.[10] Several were reprinted five or six times in the following centuries so they were clearly in demand and copies made their way all across Europe. Printed copies of the works of Buridan, Oresme, Swineshead and Heytesbury were also made available in all the main cities of Europe.[11] A copy of *De*

*Caelo et Mundo* (*Of the Heavens and the World*) by Buridan's Occamist student, Albert of Saxony, probably printed in Pavia in 1482, was one of those stored in Leonardo's chest. Albert's text was the likely source of Leonardo's knowledge of mathematical laws of motion and impetus.[12] Far from being forgotten, the philosophy of William of Occam and his followers was alive and well in the centuries following the plague, where it not only provoked lively discussion in Florentine music rooms, but had a profound influence on the cultural revolutions that ended the Middle Ages and inaugurated the modern world.

## The dark God of nominalism

Just as God creates every creature merely from His volition, so He can do with creatures whatever pleases Him merely from His volition. Hence, if someone should love God and perform all the works approved by God, still God could annihilate him without any offence. Likewise, after such works God could give the creature not eternal life but eternal punishment without offence. For God is debtor to no one.

William of Occam, Commentary of the Sentences, 1324[13]

Despite reading this passage written by William of Occam seven centuries ago many times, I am still shocked by it. God may not yet be dead,[14] but his nominalist, unknowable and omnipotent stump leaves very little purchase for mere mortals. It is disquieting today but must have found resonance in a Europe where people were still reeling from their encounter with an implacable enemy they believed had been visited on them by God: the plague bacillus. In his 2008 book *The Theological Origins of Modernity* the American philosopher and historian Michael Allen Gillespie argues that Occam's philosophy provided the spark that lit both the Renaissance and the Reformation. According to Gillespie, it was Europe's encounter with Occam's philosophy that 'turned the world on its head'.

## *Renaissance man*

Ah new people, haughty beyond measure, irreverent to so great a mother!

Petrarch, Canzone 53, st. 6

The Renaissance, with its raft of political, cultural, social and artistic changes taking place over several centuries, had as many causes as manifestations. These included the depletion of half of Europe's workforce due to the plague, which in turn prompted serfs to flee their masters to seek paid work. Without a plentiful and compliant source of peasant labour, feudalism mostly collapsed. The plague also undermined confidence in the Catholic Church since its survivors knew that a million prayers, countless confessions, and a multitude of masses had been powerless to halt the catastrophe. A new spirit of scepticism spread throughout the continent. One of the most influential sceptics was the scholar and poet Francesco Petrarca (1304–74), commonly known as Petrarch, who is often credited as being the father of the philosophy of the Italian Renaissance, humanism.

Petrarch was born in Tuscany but spent most of his early life in Avignon, overlapping with William of Occam's enforced sojourn in the city. Like Occam, he attacked the papacy as corrupt and hypocritical. Though educated in the scholastic tradition, Petrarch grew to hate the clunky logic of Aristotle, preferring instead the elegant prose of ancient Rome, particularly the lawyer, orator, writer and diplomat Cicero (106–43 BCE) who first used the Latin term *humanitas* to refer to learning and reason focusing on humanity rather than the gods. Petrarch travelled widely across Europe and spent several years in Florence where he might have been one of the 'savage dogs' in Landini's dream. Although he never mentions Occam in his writing, Petrarch must surely have known of the English scholar who had caused such a stir in his hometown.

The roots of Petrarch's humanism remain hotly debated, but

several scholars have argued that they lie in his encounter with what Gillespie calls 'the dark God of nominalism'.[15] Like the nominalists, Petrarch rejected philosophical realism and the existence of universals. He also insisted on the primacy of God's omnipotence and the unknowability of divine will. In his essay 'On His Own Ignorance and That of Many Others' he maintained that 'In this life it is impossible to know God in his fullness' and that 'nature has created nothing without strife and hatred'.[16] If God's plan for mankind is unknowable then, Petrarch concluded, man must put trust in his own creativity. He insisted that 'nothing is admirable except the soul; compared to its greatness, nothing is great'.[17] With no realist universal of *humanness* to draw upon, mankind, he insisted, must fashion its own nature. His response to the nominalist God was to invent a radically individualised humanity and he urged his fellow men to abandon the hopeless search for external validation and instead recover/seek their humanity through introspection. For Petrarch, self-examination and a creative imagination could even achieve a kind of divine status. He asked, 'What more, pray, could man, I do not say hope for, but aim at, and think of, than to be God?'[18]

The American art historian and Renaissance scholar Charles Trinkaus (1911–99) sees William of Occam's influence also in Petrarch's poetry.[19] Trinkaus argues that Petrarch seized upon the nominalist freeing of words from the tethers of Platonic realism to open up an entirely new world of flexible poetic metaphor. The American literary critic Holly Wallace Boucher[20] takes up this theme to argue that whereas earlier medieval writers, such as Dante, understood words to have 'a simple and intelligible relationship to the truth and [to be] an image of the divine order', Occam's nominalism shattered this rigid relationship. Thereafter, words could mean whatever the poet chose them to mean, or even, from a postmodernist perspective, what each individual reader chooses them to mean.[21] So, in Giovanni Boccaccio's *The Decameron*, written only thirty years after Dante's death, words had already escaped their symbolic or divine meaning to create a more naturalistic poetry appropriate for descriptions of ordinary people engaging in the normal human activities of cooking, eating, chatting, drinking, lusting, fornicating and cheating on each other. With words unencumbered by their realist anchors, they were, at last, freed to provide the metaphors that allow our common understanding of poetry:

> . . . *Look, love, what envious streaks*
> *Do lace the severing clouds in yonder East:*
> *Night's candles are burnt out, and jocund day*
> *Stands tiptoe on the misty mountain tops.*
>
> Shakespeare, *Romeo and Juliet*, iii.5

Francesco Landini was not the only artist influenced by William of Occam's nominalism. In his *The History of Art as the History of Ideas*,[22] the Viennese art historian Max Dvořák (1874–1921) argued that Occam's nominalism helped to provoke that shift of focus from a God's-eye philosophical realist perspective typical of Byzantine, medieval European as well as Islamic* painting styles towards the naturalism that is characteristic of modernity. Archetypes gave way to individuals and a rabbit in a painting could finally represent just a rabbit.

Of course, art, like everything else, did not change overnight. Symbolism, allegory and archetypes persisted in European art for several centuries, often alongside naturalist representations. Yet, this later phase of symbolism tended to be more in the nature of a secret code between the artist and the well-informed viewer, rather than an attempt to represent God's message. In his *The Social History of Art* published in 1951, Arnold Hauser (1892–1978)[23] argued that 'Realism is the expression of a static and conservative dynamic . . . [whereas] Nominalism, which claims for every particular thing a share in being, corresponds to an order of life in which even those on the lowest rung of the ladder have a chance of rising.' Nominalism, Hauser claimed, prompted a more democratic and

---

* This clash between the God's-eye view of Persian miniature art and the individualistic perspective of Renaissance Western art is strikingly depicted in Orhan Pamuk's wonderful novel *My Name is Red*.

naturalistic artistic style in which ordinary mortals were accorded the same attention as kings or saints. The shift can be seen in Leonardo da Vinci's famous painting (around 1495) of *The Last Supper* where each of the Apostles already fills the same space on the canvas as Christ, something rare in earlier paintings. The shift is even more apparent a hundred years later in Caravaggio's *The Supper at Emmaus* (painted around 1600) where the anonymous innkeeper is probably the most imposing figure in the scene.

## Humanism and Hermes

While William of Occam's nominalism was an inspiration for the pioneers of humanism, as the Renaissance developed it increasingly turned its back on the scholastic philosophers and Aristotle. The catalyst was the scholar, priest and adviser to Cosimo de' Medici, Grand Duke of Tuscany, Marsilio Ficino, who lived 1433–99. By this time, contact with the Byzantines had brought knowledge of Greek back into western Europe and Ficino translated several of Plato's works directly from the Greek into Latin. He became enamoured with Plato's philosophy and particularly his extensive writings on the nature of the soul, which he saw as a means of combating what, for him, was the dangerous tendency among the nominalists to separate philosophy from religion. Sometime between 1469 and 1474, he put together a commentary and summary of Plato's ideas under the title *Theologia Platonica* (*Platonic Theology*) with the provocative subtitle *On the Immortality of Souls*. Ficino insisted that Plato, rather than Aristotle, was Christianity's patron philosopher.

Ficino's translations and commentaries were enormously popular as Plato's lucid and poetic writing style, filled with allegories, narrative and dialogues, were a welcome release from the clunky logic of Aristotle. What particularly attracted the humanists was Plato's focus on self-discovery. The shift from Aristotle's empiricism to Platonic introspection resonated with the prevailing humanism and sparked a revival of mystical and magical philosophy known as Neoplatonism, which had flourished in the dying years of the Roman Empire. When

Cosimo de' Medici heard that some re-discovered Neoplatonist Greek texts had recently arrived in Florence he ordered Ficino to abandon further translations of Plato to focus instead on translating ancient writings by a legendary character known as Hermes Trismegistus, reputed to be an ancient Egyptian 'thrice great priest, prophet and lawmaker'.

In his introduction to what is often known as the *Hermetic corpus* or *Hermetica*, published in 1471, Ficino relates how 'At the time at which Moses was born, the astrologer Atlas flourished, brother of the natural philosopher Prometheus and grandfather of the great Mercury, whose grandson was Hermes Trismegistus . . . They say he killed Argus, ruled over the Egyptians and gave them their laws and letters.' Ficino claimed that the newly translated texts, with their mixture of philosophy, Pythagorean mysticism, alchemy, magic, mythology and astrology, were a window into an even more ancient mystical tradition that had inspired Pythagoras, Plato and the Hebrew Bible.

Despite its outlandish claims, which would have been dismissed as nonsense a hundred years earlier, hermeticism became hugely popular among humanists who had an interest in the powers of the unfettered human imagination. Hermeticism appeared to provide the missing piece to the humanist puzzle of how to turn man into a kind of creative god. The answer was magic. The Italian nobleman, and close friend of Lorenzo de' Medici, Giovanni Pico della Mirandola (1463–94), author of *Oration on the Dignity of Man*, credited as the manifesto of the Renaissance, claimed that angels helped people to fly and that his own mastery of the Hebrew Kabbalah* conferred on him the ability to speak words of magical power.[24] Indeed, hermetic philosophers claimed that the entire cosmos was a network of magical forces so that answers to any question could be read in the stars. Astrology, which had been discouraged during the scholastic period, was back in fashion, with rulers employing their own astrologers, such as the occultist

* An ancient Jewish mystical interpretation of the Bible with the aim of achieving a kind of union with God.

philosopher, reputed magician and adviser to England's Elizabeth I, John Dee (1527–1608).

Alchemy also made a revival, particularly as a recipe book for magical potions that could cure all ills. The Swiss-German Philippus Aureolus Theophrastus Bombastus von Hohenheim (1493–1541), better known as Paracelsus, claimed that sickness was a consequence of disharmony with the cosmos that could be balanced by magical potions whose composition was written in the stars. Ficino insisted that 'the occult virtues of things . . . come not from an elementary nature but from a celestial one'. So, in stark contrast to the nominalist quest for simplicity, the Renaissance humanists invented a plethora of magical, mystical and occult entities that went way beyond necessity.

Nevertheless, despite its mystical leanings, humanism did at least retain an interest in the world and how it could be manipulated. In northern Europe, Occam's nominalism was stirring a very different intellectual potion.

## The northern route to the Reformation

The Dutch Occamist scholar Marsilius of Inghen (1340–96) was a key figure in the transmission of the *via moderna* to Europe's northern universities. He had studied with Jean Buridan and Nicole Oresme in Paris and went on to teach there between 1362 and 1378. From Paris, he travelled to Heidelberg, helping to found its university in 1386, with a strongly nominalist curriculum. He wrote prolifically, producing commentaries on Aristotle's *Physics*, *Metaphysics*, *On the Soul* and *On Generation and Corruption*, as well as several texts on logic including *Various Questions on the Old and New Logic* (nominalism). These texts were widely copied and disseminated to universities and libraries in Prague, Kraków, Heidelberg, Erfurt, Basle and Freiburg, helping to spread the *via moderna* throughout the universities of northern Europe. Gabriel Biel (1420–96), one of Marsilius's most influential students, introduced his own brand of Occamist philosophy to one of Germany's oldest and most influential academic institutions, the University of Erfurt, such that, by the end of the fifteenth century, nominalists dominated all but one university in Germany.[25]

In 1501, thirty years after Ficino published his *Hermetica* in Florence, the University of Erfurt accepted a young student of philosophy and law named Martin Luther. Luther had been born in 1483 in Eisleben, Saxony, a region of the Holy Roman Empire at that time. His father had been a peasant but, after taking up mining, had prospered sufficiently to send his son to school and then to the University of Erfurt. Biel had died five years before Luther arrived in Erfurt, so it was his pupils, Johannes Nathan and Bartholomaeus Arnoldi von Usingen, who introduced Luther to the writings of William of Occam.

Luther's formative university years were hugely influenced by Occam, whom he later referred to as his 'dear master', insisting that 'Occam alone understood the logic'.[26] As a youth, Luther accepted the nominalist rejection of universals and embraced its frighteningly omnipotent and unknowable God. He later wrote that he lived in terror of an unknowable but wrathful God, even to the extent of being reluctant to hold the communion host in the Mass.

In July 1505, Luther left Erfurt's university to join its Augustine monastery where he remained for several years before accepting a post to teach theology at the University of Wittenberg. By this time, humanism had spread northward from Italy, to France, Germany, England and the Low Countries, where its most influential convert was that other intellectual pillar of the Reformation, Erasmus (1466–1536). The illegitimate son of a priest and physician's daughter, Erasmus had been orphaned and brought up by guardians who sent him to a school run by the Brethren of the Common Life, a religious community established in the fourteenth century on the principles of apostolic poverty. Aged twenty-five, he entered an Augustinian monastery and was ordained as a priest; but he grew to loathe the monastic life so travelled to Paris to study for a bachelor's degree in divinity. There he developed his own brand of humanism, less egocentric and elitist and more spiritual than its Italian counterpart. Erasmus also placed more trust in the scriptures than the Italian nominalists, emphasising the humanity of Jesus as a counterweight to the unknowable nominalist God. Nevertheless, like the southern humanists, Erasmus believed that man can only come to God through self-knowledge. This idea was anathema to Luther.

Like St Augustine, Luther considered man to be hardly worth knowing because humanity was mostly a heap of depravity. With many echoes of William of Occam's nominalism, he claimed that the 'natural condition of the world is chaos and upheaval' because 'in His own nature God is immense and incomprehensible and infinite . . . He is intolerable, a hidden God, as it says in the scriptures "No man may see me and live".'[27] Luther rejected the southern humanists' flight to creativity to escape the frightening nominalist God. He also rejected Erasmus's gentler humanism and particularly his claim that God could be reached through self-knowledge and reason. He insisted instead that God's omnipotence was incompatible with human free will. For Luther, the fate of every human, whether they were destined for heaven or hell, was decided by God before they were even born. The devout were not devout because they chose to follow God; instead they followed God because God, in his unknowable wisdom, had made them that way. Faith, to Luther, was not a choice but a mark of God's favour.

## Science and cultural revolution

In many ways, there were only bad choices available to any believer confronted by the unknowable omnipotent God of William of Occam and his nominalist followers. The first was to adopt the head-in-the-sand approach by carrying on as before regardless of reason and putting one's trust, not in reason, but in the authority of the Church. This, perhaps unsurprisingly, was the approach eventually adopted by the Catholic Church, which rejected its flirtation with nominalism to return to philosophical realism and the benevolent God of Thomas Aquinas. It maintains that position to this day.

The second and third options both accepted divine omnipotence but pushed the dark God of the nominalists to the periphery. The downside of this approach was that it left mankind in a potentially meaningless universe. The humanist response, the second option available and the one adopted by Petrarch, the Italian humanists and, to a lesser extent, Erasmus, was to fill the gap in meaning with man, raising him to the status of demigod. The third option, the one taken by Luther, was to turn to the scriptures as the arbiter of truth and the source of meaning in the world.

These were all poor options. Each was a kind of fudge that took onboard the nominalist God but then, in one way or another, rejected Him. Thereafter, neither the followers of the Renaissance nor the Reformation were likely to keep William of Occam close to their hearts. From their perspective, he was like the childhood friend who tells you that Santa Claus didn't exist. You can't ignore your new knowledge but, in your heart of hearts, you know that the world has lost some of its enchantment. And you no longer want to hang around with the ruthlessly logical friend who has exposed the truth.

Nevertheless, I believe that the biggest impact of Occam's ideas was not on philosophy or theology, but on the science that emerged from this intellectual maelstrom. In many ways, the sparser philosophical base of the Lutherans, with their more enthusiastic adoption of *via moderna* principles, including its separation of science from theology, was closer to the empiricism of modern science. For example, Luther's teacher at Erfurt, Bartholomaeus Arnoldi,[28] had taught that science should be tested through experiment and reason; whereas theology can only be revealed through the scriptures. The Lutherans also adopted a healthy scepticism towards the products of the human imagination, including the mystical and occult ruminations of the southern humanists. Yet while the northern humanists were largely uninterested in science, preferring to search for the truth in the scriptures rather than the world, their indifference did leave cultural ground where the seeds of science could germinate.

It is perhaps ironic therefore that it was Renaissance humanism in the shape of Italian polymath Leonardo da Vinci that produced the greatest scientist of the age, as revealed in his extraordinary *Codices*. His unique synthesis of art, technology and science could have been the springboard for a sixteenth-century scientific revolution, particularly as he also held a healthy scepticism towards the pseudosciences of his age, such as astrology and alchemy. However, Leonardo never published his ideas and, as far as we know, no one, except Leonardo, read his *Codices* in his lifetime. After his death, the keepers of Leonardo's *Codices* admired them for their art not their science. Lacking Leonardo's intellectual rigour and fascination with the natural world, and with no respect for Occam's razor, the southern humanists mostly embraced the occult.

With the Lutherans largely dismissive of science and the humanists practising their spells, science might, once again, have stagnated. However, an unlikely collaboration between the humanist canon of the cathedral in Catholic Kraków and a scholar educated in Lutheran Wittenberg found a simple path out of this predicament.

# PART II

## The Unlocking

# 7

# The Heliocentric but Hermetic Cosmos

Three years before Leonardo's death in 1519, Georg Joachim Iserin was born in the town of Feldkirch in what is now Austria. His father was a well-to-do doctor who possessed a fine library and he sent his son to the local grammar school where Georg learned Latin and the liberal arts of grammar, rhetoric and logic. When Georg was fourteen, his father was tried for theft, larceny and sorcery. He was convicted and sentenced to death. The family's name was struck out of existence so Georg's mother reverted to her Italian maiden name of Thomasina de Porris (of leeks). Georg became Georg Joachim de Porris but, as the boy did not consider himself Italian, he translated his name into German as Georg Joachim von Lauchen. Later he added the name of Rheticus after the Roman province of Rhaetia in which he had been born. It is by this name that he is usually known today.

Rheticus's mother was well connected and wealthy so the boy continued his education under the tutelage of a friend of Erasmus, Oswald Myconius. In the autumn of 1531 Rheticus returned to Feldkirch where he struck up a lifelong friendship with the physician who had taken over his father's practice, Achilles Gasser. As well as being the town doctor, Gasser was a well-known humanist scholar with interests in history, mathematics, astronomy, astrology and philosophy.

In 1533, aged nineteen and armed with letters of introduction from Gasser, Rheticus travelled 400 miles north-east to the university town of Wittenberg where Martin Luther held the chair in theology. A decade earlier, the fiery young monk had made himself the most influential member of the Protestant German establishment by using his pulpit to condemn the German Peasants' Revolt of 1524 on the scriptural grounds of Jesus's advice to 'Render unto Caesar the things which are Caesar's, and unto God the things that are God's'.* The revolt was brutally crushed with the slaughter of around 100,000 poorly armed peasants. It transformed Luther's reputation from that of an irksome rebel to a pillar of the German establishment.

Various shades of Lutheran Protestantism quickly swept across Germany and into Switzerland, France and the Nordic countries before crossing the Channel to England. Thereafter, the European continent split along a roughly north–south axis with the predominantly Lutheran German states in the north and the humanist-influenced Catholic countries, such as Spain, France and Italy, in the south. The north–south divide was mirrored by a philosophical dispute on the nature of the human will. The humanists placed human will, inspired by human creativity, as central to human nature. In his *On the Freedom of the Will*, published in 1524, Erasmus had argued that, despite God's omnipotence, mankind nevertheless possessed free will as a gift from God. Luther was unconvinced and, in 1525 – nine years before Rheticus arrived at Wittenberg – replied with one of the most influential texts of the Reformation. In *On the Bondage of the Will* Luther reiterated that mankind is a slave to God's will, writing that 'God . . . knows no limit but omnipotence' and that anyone who thinks otherwise is not a Christian.

With absolute trust in the Bible providing the only escape from hellfire, many of Luther's followers attacked the very notion of education other than study of the scriptures. Luther was not so fundamentalist. In 1518 he appointed a brilliant German humanist scholar, Philip Melanchthon, to the chair of Greek at Wittenberg. Melanchthon became Luther's most trusted disciple and adviser who balanced his mentor's hot-blooded temper and earthy language with a quiet manner and a willingness to persuade rather than bully. His influence helped to soften the edges of German Protestantism

* Matthew 22:21.

so that, although it retained Luther's grim vision of human predestination and the central importance of the scriptures, it nevertheless developed a humanist tolerance for learning outside of theology and an interest in the world beyond the Bible.

Melanchthon welcomed the young Rheticus into the humanist-inclined intellectual circle that he had established at Wittenberg and, in 1536, appointed him to teach mathematics and astronomy. By this time, rumours of a radical new model of the cosmos with a moving earth that circled the sun was sweeping through the universities of Europe. Most of his Wittenberg colleagues treated the reports with a mixture of disbelief, derision and humour but Rheticus was intrigued.

## The earth moves for the mystic astronomer

Born in 1473, forty-one years before Rheticus, in the city of Toruń in the Warmia region of what is now northern Poland, Copernicus attended the University of Kraków where he received the typical scholastic education in the seven liberal arts, with its emphasis on Aristotle and his Arab and Christian commentators. There he learned about Aristotle's physical model of the cosmos, together with Ptolemy's geocentric model and the mathematical systems used to calculate the heavenly motions. This was also the peak of the *via moderna* in European universities, including Kraków.[1] According to the Polish historian Władysława Tatarkiewicza, 'The *via moderna* had its adherents from the very beginning at Cracow. More specifically, in physics, logic and ethics, Terminism [nominalism] prevailed here under the dominant influence of Jean Buridan.' So there is no doubt that Copernicus was exposed to the ideas of William of Occam, his razor and his followers while at Kraków.

In 1496, aged twenty-three, Copernicus left Kraków without graduating and travelled to Italy to study canon law at Italy's oldest university in Bologna, where the Occamist scholar Alessandro Achillini (1463–1512) was professor of philosophy and medicine.

Two years earlier another Occamist, Marcus de Benevento, had published Occam's commentary on Aristotle's *Physics* in Bologna and dedicated the edition to Achillini. De Benevento went on to publish three more works by Occam in Bologna, culminating with his *Summa Logicae* in 1498,[2] the same year that Copernicus arrived in the city. Also that year, de Benevento published a collection of the works of the nominalist Albert of Saxony 'in honour of Brother William of Occam'. Bologna was clearly awash with the works of William of Occam and the *via moderna* scholars during Copernicus's education in the city.

Although Copernicus came to Bologna to study canon law, it seems it was there that his interests shifted decisively towards astronomy, even leading him to conduct some of his first astronomical observations in the city. After a brief spell back in Warmia, he returned to Italy in 1501, but to the University of Padua to study medicine. By this time, the centre for European intellectual life had shifted from Oxford and Paris towards Renaissance Italy and particularly Padua. It was probably in Padua that Copernicus became infatuated with the city's predominant Neoplatonist philosophy, mysticism and Hellenic focus, which later dominated his intellectual life.

Copernicus eventually returned to Warmia to take up his canonry in Frauenburg (now Frombork in Poland) sometime in 1503 when he was thirty. His duties did not appear to be too onerous as Copernicus continued his Hellenistic interests, translating Greek poetry into Latin. He also attempted to apply humanist principles to his other great interest, astronomy, and particularly the Ptolemaic system, but there he encountered a problem. Instead of the Neoplatonist perfection that he expected in the heavens, he found a confusion of epicycles, equants and deferents. He later wrote that

> I was impelled to consider a different system of calculating the motions of the universe's spheres for no other reason than the realization that the astronomer's . . . experience was just like someone taking from various places hands, feet, a head, and other pieces [to make] a monster rather than a man . . . from them.

The reference to the parts of the human body is interesting. This was about fifteen years after Leonardo da Vinci had drawn his famous *Vitruvian Man*. The work was probably inspired by the humanist Marsilio Ficino's proclamation that 'Man is the most perfect animal . . . he is connected to the most perfect things, i.e. divine ones.'[3] Copernicus appears to be contrasting Ptolemy's monstrous astronomical system with the humanist dream expressed in Leonardo's painting of an orderly mathematical universe with divinely proportioned Man at its centre. Copernicus became convinced that with the help of mathematics, he could reassemble the cadaver to reveal a more harmonious heaven. He wrote that, 'having become aware of these defects, I often considered whether . . . it could be solved with fewer and much simpler constructions than were formerly used.'

Like his *via moderna* predecessors Copernicus allowed himself to be guided by Occam's razor. On the relativistic observer principle that Occam had highlighted, he argued that accepting that the earth rotates every day, rather than the sun, moon, planets and stars, provides a much simpler cosmos.

## Making the world less arbitrary

Simplifications often deliver unexpected bonuses. Allowing the earth to spin to still the stars provided Copernicus with the bonus of eliminating five of Ptolemy's planetary epicycles. These had essentially (but of course unknowingly) corrected for a stationary earth in Ptolemy's model, by relativistically transferring terrestrial rotation to Ptolemy's planetary epicycles. Another bonus was clarity. With a less cluttered model Copernicus was able to see a way to remove more epicycles by his second and even more revolutionary move of shifting the system's centre from the earth to the sun.

Copernicus was not the first scholar to place the sun at the centre of the cosmos. Aristarchus of Samos had, around 250 BCE, proposed a heliocentric system, but the idea had fallen out of favour in antiquity as it clashed with Aristotle's insistence that all heavy objects, including planets, either fall towards, or orbit around, the centre of the earth. However, with the authority of at least one of the ancients behind him, Copernicus dared to reinstate the sun at the centre of all heavenly rotations. He was astonished to discover that the move provided 'a marvellous symmetry of the universe, and an established harmonious linkage between the motions of the spheres and their size, such as can be found in no other way'.[4]

Here, Copernicus had stumbled upon a phenomenally important signature of a successful simplification of a system: elimination of its arbitrary features. Ptolemy's geocentric system could not explain why Mercury and Venus are always seen to be closest to the sun at sunrise and sunset. He accounted for this observation by adding an arbitrary rule to his model: Venus and Mercury just happen to orbit the earth (on their own epicycles) while keeping close to the rotation of the orbiting sun (Figure 9a).

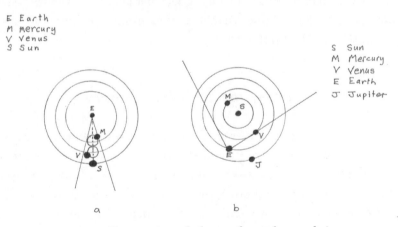

FIGURE 9: Perspective of planets from the earth in a
(a) geocentric or (b) heliocentric system.

However, when Copernicus shifted the sun to the centre of planetary rotations, then he was free to move Venus and Mercury to a position between the earth and the sun so that they became the inner planets. In this position, their closeness to the sun in the sky is simply due to their closeness to the sun in reality (Figure 9b).

In this way, an arbitrary feature of the complex model becomes an inevitable consequence of its simple alternative.

Another unanticipated bonus of Copernicus's heliocentric system was that it finally made sense of the retrograde motion of the planets Mars, Jupiter and Saturn. These normally travel east to west across the sky, along with the sun and stars, but sometimes reverse their motion to travel west to east for many weeks before turning around and heading east again (Figure 4). Copernicus noticed that the wandering stars that performed these pirouettes were all, in his heliocentric system, outer planets further away from the sun than the earth. Ptolemy had incorporated their retrograde twists by introducing more epicycles whose only justification was to make his model fit the data. With the sun in the centre, those epicycles vanish as they become an inevitable consequence of the earth catching up with an outer planet in its orbit and then overtaking it. We experience something similar when we pass a slower-moving vehicle on the highway. When it is ahead, it appears to be moving against the surrounding landscape and in the same forward direction as ourselves. However, as we pass the vehicle it initially appears to switch its direction of travel to be moving backwards against the background, until, after we pass and view it in our rear-view mirror, it appears to be again moving forward. Retrograde motion in the heavens is the same kind of optical illusion but seen from the earth when it overtakes a slower-orbiting outer planet, such as Mars. Another arbitrary feature of the complex model becomes a necessary consequence of its simpler alternative.

With these successes, Copernicus could have provided a much simpler model of the cosmos . . . if he had stopped there. Unfortunately, heliocentricity had not eliminated that vexing equant that had prompted him to search for a new cosmic model. To eliminate this, while retaining circular orbits, Copernicus had to resort to the Ptolemaic device of introducing several new epicycles, effectively adding back much of the complexity that heliocentricity had eliminated.

Despite this setback, in 1514, the year Rheticus was born, Copernicus posted a short astronomical treatise entitled *Commentariolus* (*Little Commentary*) to a select group of European scholars. His paper created a ripple of curiosity that even reached Rome where it received a positive reception. In 1517, he received a letter from the Pope's secretary, Cardinal Nicholas Schoenburg, urging Copernicus 'to communicate your discovery to the learned world'. Copernicus initially appeared to take the cardinal's advice to heart as he began work on an extensive account of his heliocentric model in *De revolutionibus orbium coelestium* (*On the Revolutions of the Heavenly Spheres*). Yet he never attempted to publish his manuscript nor, as far as we know, allowed anyone to read it.

## The earth doesn't move Luther

In the following decades, Copernicus's treatise describing his heliocentric model of the cosmos was mostly ignored or derided; but it did circulate widely, reaching even the centre of the Lutheran world at Wittenberg. One infamous after-dinner remark by Martin Luther describes 'that fool who wishes to turn the whole of astronomy upside down. Even in these things that are thrown into disorder believe the Holy Scriptures, for Joshua commanded the sun to stand still, and not the earth.'[5]

Despite Luther's scorn, Rheticus was intrigued and requested permission to visit the elderly canon in Catholic Poland. It seems unlikely that Melanchthon would have granted the young scholar leave to visit Copernicus but, around 1538, a scholarly prank put Rheticus in a potentially dangerous position. Simon Lemnius, a member of the Wittenberg humanist circle that now included Rheticus, wrote a series of satirical and erotic verses in the style of the Roman poet Ovid, but featuring Luther. He even sold his anthologies in front of the Wittenberg church door on which, according to legend, Luther had twenty years earlier nailed his Ninety-Five Theses against the Catholic Church.

Not known for his sense of humour, Luther responded by accusing the poet of slander, prompting Lemnius to flee the city. The following Sunday, Luther preached a sermon warning that the author would likely lose his head. In September of that year Luther composed his own poem entitled 'A Dysentery on Lemmy the Shit-Poet'.[6] From exile, Lemnius replied with another verse including the lines

> *Where formally your crooked mouth spewed madness*
> *It's now your arse from which you vent your spleen.*

He also published a more restrained *Apologia* in which he claimed to represent the more moderate elements of German Protestantism, citing both Melanchthon and Rheticus as his supporters and allies. Melanchthon was powerful enough to brush off any guilt by association but Rheticus was vulnerable to becoming a scapegoat for Luther's wrath. Perhaps with this danger in mind, in October 1538, Melanchthon gave permission for Rheticus to leave Wittenberg to visit 'that fool who wishes to turn the whole of astronomy upside down'.

Rheticus took his time. He first visited many of the most eminent astronomers in Europe, before diverting to his hometown of Feldkirch to present a printed copy of Sacrobosco's astronomy textbook *De sphaera mundi* to his mentor Achilles Gasser. He eventually arrived in Frauenburg in May 1539. It was not an impressive sight. A ramshackle town on the southern fringe of the lagoon where the River Vistula enters the Baltic, Frauenburg boasted a small harbour where fishermen moored flat-bottomed boats from which they fished eels out of the lagoon. Above the town loomed its massive and rather ugly red-brick cathedral. Copernicus called the place 'this remotest corner of the Earth'. By this time, the canon had been sitting on his revolutionary heliocentric hypothesis for more than thirty years.

## How to be right when you're wrong

As well as his enthusiasm for heliocentricity, Rheticus brought gifts of mathematics and astronomy books, including a newly printed edition of Ptolemy's *Elements*. Copernicus was delighted. After years of academic obscurity, he finally had a student. The young scholar had planned to stay only a few weeks but he soon became so devoted to the elderly Copernicus, whom he constantly referred to as 'my teacher', that he remained for two years, helping to revise and edit his great book on the heliocentric system.

Nonetheless, Copernicus would only agree to Rheticus writing his own account of the unpublished 'Revolutions'. This text became *Narratio Prima* (*The First Account*), which describes the author as 'a certain youth most zealous for mathematics', and in which Copernicus is referred to only as the 'Teacher' or 'the learned Dr Nicolaus of Torun'. Undeterred, Rheticus headed for Danzig to find a publisher.

*Narratio Prima* was published in 1540. Rheticus sent copies to everyone of influence he knew, including the astronomer Johannes Schöner and his mentor back in Feldkirch, Achilles Gasser. Schöner passed a copy to the Nuremberg-based printer Johannes Petreius, who responded by insisting that he 'consider[ed] it a glorious treasure' and urged Copernicus to publish a full account. The positive reception at last persuaded the reticent canon to publish his *On the Revolutions of the Heavenly Orbs*.

Rheticus returned to Wittenberg to resume his lectures, but found time to lobby both publishers and political forces in support of his ambition to have *Revolutions* published. In 1541 he was back in Frauenburg where he copied and edited Copernicus's book. In the spring of 1542, Rheticus set off for Petreius's presses in Nuremberg with the precious text in his bag. However, by this time Rheticus had been recruited to a new post at the University of Leibniz and had to go there to teach his classes. He entrusted supervision of the printing process to the Nuremberg-based Lutheran theologian, and mathematician, Andreas Osiander.

## *The infamous preface*

*On the Revolutions of the Heavenly Orbs* was finally published in 1543. It described Copernicus's heliocentric cosmos with its key innovation of allowing the earth to move, rotating on its own axis every day and orbiting the sun every year. Even more significantly the earth was demoted from its prime position as the centre of the cosmos to being merely the third of six planets orbiting the sun. The terrestrial world was no longer special.

However, in what is generally held to be one of the greatest scandals in the history of science, an anonymous preface was added to Copernicus's book addressed 'to the reader concerning the hypotheses of this work'. The preface begins by stating that the ideas in the book should not be taken too seriously: 'For these hypotheses need not be true or even probable. On the contrary, if they provide a calculus consistent with the observations, that alone is sufficient.' In other words, Copernicus's thesis was just another example of 'sky geometry', with no more claim on reality than Ptolemy's epicycles. He concludes with the assessment that 'So far as hypotheses are concerned, let no one expect anything certain from astronomy, which cannot furnish it, lest he accept as truth ideas conceived for another purpose, and depart from this study a greater fool than when he entered it. Farewell.'

A copy of the first print run of the book was rushed to Canon Copernicus's bedside. He died the same day of 24 May 1543, his death, it is said, hastened on reading the infamous preface. For many years its author remained unknown. It was Johannes Kepler who later uncovered the truth that it had been written by Andreas Osiander.

The publication of *On the Revolutions of the Heavenly Orbs* is often considered to be a landmark in the history of science, even marking the birth of modern science. If that is true, hardly anyone noticed. The first edition of 400 copies failed to sell out and it was more than twenty years before a second edition was printed. As far as we know, no eminent astronomer adopted the Copernican system, preferring instead the tried and tested Ptolemaic methods. Crucially, Copernicus's treatise had not been able to cite any proof of his heliocentric system, and it was no more accurate at making astronomical predictions than Ptolemy's geocentric system. Each provided an accuracy of prediction of about one degree of arc.*

---

* An angular distance of separation in the sky roughly corresponding to the thickness of your little finger if you hold it as far as you can away from you.

However, you will remember that Copernicus did not place the sun at the centre of his system to make more accurate predictions. Instead, it was his horror at the Byzantine complexity of the Ptolemaic model that motivated his search for a system 'with fewer and much simpler constructions than were formerly used'. Did he deliver on his aim to construct a simpler model? On a simple circle count, no. Counting the final number of circles in either the Ptolemaic or Copernican system isn't straightforward as neither produced an overall model of their systems, only diagrams of parts and recipes for calculating planetary positions. The consensus is that both systems have somewhere between twenty and eighty circles depending on what counts as a circle.[7]

Nevertheless, despite their similar circle counts, Copernicus was convinced that his system made better sense than Ptolemy's because it was simpler. In his *Revolutions* he wrote that 'This should be admitted, I believe, in preference to perplexing the mind with an almost infinite multitude of spheres, as must be done by those who kept the earth in the middle of the universe.' He goes on to argue that 'On the contrary, we should rather heed the wisdom of nature. Just as it especially avoids producing anything super-fluous or useless, so it frequently prefers to endow a single thing with many effects.'[8] Crucially however in the *Revolutions*, Copernicus's claim for simplicity is not based on a circle count. Instead, he cites features of his model that eliminated arbitrary complications of Ptolemy's. These included the elimination of terrestrial daily cycles in the orbits of the heavenly bodies, the removal of cycles needed to account for retrograde motion and an unambiguous ordering of the planets. It was these features rather than any circle count that according to Owen Gingerich, professor of astronomy and the history of science at Harvard University, convinced Copernicus that heliocentricity delivered a 'new cosmological vision, a grand aesthetic view of the structure of the Universe'.[9] Copernicus's trust in simplicity was well rewarded. With Occam's razor in hand, even mystics like Copernicus could find the path towards modern science.

# 8

## Breaking the Spheres

Although Copernicus's system had the right centre, it was nevertheless cluttered. Moreover, those crystal spheres still hung in the heavens making the cosmos finite, bounded by the outermost celestial sphere.

### *The heavenly count*

Tycho Brahe (1546–1601) was born three years after Copernicus's death, in Knutstorp, Denmark to members of the Danish nobility and the elite Rigstaad or Council of the Realm. The infant boy was abducted by his even wealthier and more powerful uncle, Jørgen, on the grounds that he and his wife were childless. So Tycho spent his childhood at his uncle's ancestral seat at the family castle at Tosterup in northern Denmark. He went on to study at the University of Copenhagen, which was, at that time, dominated by Melanchthon's humanism and particularly his emphasis on teaching science alongside theology and scripture. While in Copenhagen, Brahe's interests shifted towards mathematics, astronomy and astrology. He was hugely impressed by the ability of Ptolemy's geocentric sky geometry to predict a solar eclipse that he had himself observed in 1560; yet he found it irksome that the prediction had been a day out.[1] Brahe realised that this discrepancy must be a consequence of error, either in Ptolemy's model or its astronomical observations. Lacking the mathematical skill to tackle Ptolemy's geometry, he determined to devote his life to building better astronomical instruments capable of delivering more accurate predictions.

FIGURE 10: Portrait of Tycho Brahe.

In 1566, Brahe moved to Germany to study medicine at the University of Rostock. It wasn't long before he lost both his university place and a part of his nose during a duel with a fellow student. A prosthetic silver nose provided a solution to his disfigurement; but the duel convinced Brahe that his future did not lie in academia. Instead, he spent the following years touring his astronomical instruments around the royal courts of Europe. As well as providing intellectual entertainment to European aristocrats, Brahe also sought to find a benefactor for his ambition to build a state-of-the-art observatory.

In 1570, Brahe arrived in Augsburg in Bavaria where he managed to persuade the local alderman, Paul Hainzel, to sponsor the building of a huge astronomical quadrant, essentially a graduated quarter of a circle mounted to measure the altitude of celestial objects above the horizon. The main bulk of the new instrument consisted of an oak arch measuring five and a half metres in diameter and so heavy that it took forty men to carry it into place. It delivered unrivalled accuracy such that, Brahe claimed, had 'hardly ever [been] attained by our predecessors'.[2]

Brahe left Augsburg later that year to return to Knutstorp where his father, Otte, had fallen ill. His father died in May the following year, leaving the twenty-four-year-old Tycho as lord of a considerable estate with annual incomes from 200 farms, 25 cottages and 5 mills as well as the manorial production and seigneurial rights (essentially feudal taxes) of Knutstorp. Tycho built a new observatory, including an improved sextant, an instrument similar to a quadrant but only a sixth rather than a quarter of a circle and thereby smaller and more portable. Not content with astronomy, he also built his own laboratory utilising the services of local glass-workers to construct a variety of vessels with which to pursue the secretive art of alchemy.

## The heavenly mystic

The year 1571 was an auspicious one in astronomy as it marked not only the building of Brahe's observatory but also the birth of the astronomer who would finally make sense of the celestial confusion. Johannes Kepler was born (when Brahe was twenty-five) in the small town of Weil der Stadt in Swabia, south-west Germany. His father was a mercenary soldier and his mother the daughter of an innkeeper. In his brutally honest autobiographical account Kepler describes his childhood with his mercenary father: 'a man inflexible, quarrelsome, and doomed to a bad end'. The father left home for the last time when Johannes was five. He is believed to have died in the Dutch War of Independence. Johannes was hardly any kinder on his mother whom he described as 'small, thin, swarthy, gossiping and quarrelsome, of a bad disposition'.

Kepler's autobiography continues with his education at the local school and then at a nearby Lutheran seminary where 'During these two years [aged fourteen and fifteen] I suffered continually from skin ailments, often severe sores, often from the scabs of chronic putrid wounds in my feet which healed badly and kept breaking out again.' His school days appeared to be largely friendless; he records:

> February, 1586 . . . I have often incensed everyone against me through my own fault: at Adelberg it was my treachery . . . My friend Jaeger betrayed my trust: he lied to me and squandered most of my money. I turned to hatred and exercised it in angry letters during the course of two years.

Despite all this Kepler was awarded a scholarship to the University of Tübingen, where he studied to become a priest. From there he continues his unflattering self-portrait: 'That man [Kepler] . . . His appearance is that of a lap-dog . . . He continually sought the goodwill of others . . . he is malicious and bites people with his sarcasms.'

Kepler arrived in Tübingen in 1589. The university had been founded in 1477, a little over a hundred years after William of Occam's death, by Gabriel Biel, the Occamist philosopher described as an 'articulate spokesman of the *via moderna* and . . . a discerning user of nominalism'.[3] Biel's university became a centre of the *via moderna* in Germany so there is little doubt that in Tübingen Kepler would have been exposed to both Occam's ideas and his razor. Moreover, by the sixteenth century, Tübingen had also become a hotbed for nominalist-inspired Lutheranism, leavened by a sprinkling of Melanchthon's northern European brand of humanism. According to the theologian and historian Charlotte Methuen, it was Tübingen's peculiar blend of nominalist-inspired empiricism twinned with the mysticism and creativity of humanism that inspired Kepler's revolutionary approach to unravelling the motions in the heavens.[4] As Kepler himself wrote in 1598: 'I am a Lutheran astrologer, throwing away the nonsense and keeping the kernel.'[5] Kepler's teacher at Tübingen, Michael Mästlin, had also been a crucial influence, as he owned one of the few circulating copies of Copernicus's *On the Revolutions of the Heavenly Orbs*.

## *The new star*

In 1572, the year following Kepler's birth, Tycho Brahe was returning from his laboratory in Knutstorp when he happened to glance up at the sky and was astonished to spot a new star. He was so incredulous that he asked a group of passing peasants to corroborate the sighting. The new star's position was outside of the band of the Zodiac along which the planets travelled, so it did not appear to be a new planet. It could be a comet, a kind of wandering star known since ancient times. Yet several nights of observation convinced Brahe that the new star did not wander but instead rotated along an orderly circular path around the earth every day, like any fixed star.

Today we know Brahe had been lucky enough to witness the appearance of a rare supernova, or exploding star, now known as SN1572. Yet its appearance in 1572, so bright that for some weeks it could be seen in broad daylight, sent a shock wave through Europe. According to theologians, God had pinned each star to the celestial sphere on the fourth day of creation, where they would remain, unchanged, until the end of time. There were a few famous exceptions, such as the appearance of the star that guided the three Magi to Bethlehem and the birth of Jesus. Yet that had been a momentous moment for the Christian world, the birth of a God. What did the appearance of a new star foretell? Printing presses were soon churning out apocalyptical pamphlets warning that the star heralded the Second Coming and end of the world. To avoid this unpalatable conclusion, most astronomers opted for the possibility that the new star was not really a star at all but some other bright object, such as a comet, that had intruded into the terrestrial space between the inconstant terrestrial realm and the unchangeable heavens.

Brahe had just completed construction of his new sextant capable of settling the matter. A sextant works through the measurement of parallax, whereby objects (such as your finger held in front of your nose) appear at different positions against the background when viewed from two different points. Parallax diminishes as objects recede further away from our eye, so measuring the degree of parallax provides a means, known since ancient times, of estimating the distance to any object. Earlier astronomers could detect a small degree of parallax for the moon, but no parallax had ever been detected for any star. The Danish astronomer's instruments were the most sensitive in the world and could easily detect the moon's parallax. Yet Tycho found no parallax for the new star. It not only moved with the celestial sphere but appeared to be as distant. It was a bright stain on the supposedly incorruptible walls of God's heaven.

In 1573, Tycho put the new star aside and began working on an astronomical and astrological almanac for the year. He travelled with his almanac manuscript to Copenhagen with a view to finding a publisher, but at a dinner party with old college friends, including the humanist scholar Johannes Pratensis, he was astonished to discover that no one had even noticed the new star. They only became convinced of its existence after Tycho dragged them out into the cold winter night. Pratensis insisted that Tycho forget the almanac and instead focus on publishing his observations.

Tycho was reluctant. Perhaps, as an aristocrat, he considered scholarly work unworthy of his rank. Like Copernicus, he was also probably reluctant to offend the Church. Pratensis persevered, even sending Tycho copies of several rival published accounts of the new star in the hope that their incorrect deduction that it was a comet would spur him into action. The tactic worked and Tycho eventually wrote up his observations in a small booklet called *De Stella Nova* published in May 1573.

In *De Stella Nova* Tycho argued that the complete absence of parallax ruled out an explanation of the new star as any kind of fiery meteor or comet, and that its motion and vast distance from the earth, 'beyond the eighth sphere', could only be explained if it was fixed to the furthest, celestial sphere. The new star, Tycho insisted, must be a sign from God, likely portending warfare, plague, rebellion, captivity of princes, or other catastrophes.

*De Stella Nova* was a huge success and made Tycho the most famous astronomer in Europe. He was offered, and accepted, a post at the University of Copenhagen. However, he soon grew weary of the burden of teaching and made plans to leave Denmark, perhaps to establish himself in Germany or Switzerland. When King Frederick II of Denmark heard of these preparations, he sent an emissary to instruct Tycho to attend the king at his nearby hunting lodge. There, King Frederick offered Brahe a gift of the island of Hven, equipped with a castle and funds to build the world's greatest (though still naked eye) observatory. Brahe accepted, naming his observatory Uraniborg, and moved there in 1576.

On the island, Tycho set about building a grand house on a plan that would reflect the proportions of heavenly harmony. This was of course hugely expensive but, as lord of the island, he could demand two days of labour without pay each week from its feudal peasants. Tycho also began building his observatory, gathering and commissioning the most sophisticated astronomical instruments that the world had seen. These included a clock capable of displaying not only hours and minutes, but also seconds: a big innovation in 1577 and essential for accurate astronomy. He also built several quadrants and an armillary sphere, which comprised concentric metal rings that model the celestial spheres and their rotations (Figure 11). Most impressive of all was a brass celestial globe five feet in diameter, on which, over the course of the next twenty years, Brahe engraved the precise positions of the fixed stars.

In November 1577, Brahe enjoyed another bit of astronomical luck when an actual comet appeared in the skies above Europe. Although known since ancient times, Aristotle had thought they travelled below the moon so that they did not disturb the expected timelessness of the heavens. Brahe again used the principle of parallax to measure the comet's distance and concluded that it was more distant than the moon and within the supposedly inviolable heavens. Even more remarkably, Brahe's astronomical measurements demonstrated that, during its traverse across the sky, the comet passed unhindered through the sphere that was believed to carry Venus around its orbit. So after first tarnishing the incorruptibility of the heavens with his new star, Tycho Brahe's cometary observations had inadvertently shattered its celestial crystal.

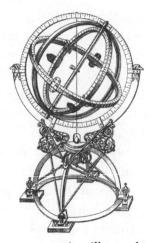

FIGURE 11: Armillary sphere.

During the 1570s Brahe devised his own new cosmological system, which was essentially a compromise between the Ptolemaic and Copernican. First he admitted that the heliocentric system 'expertly and completely circumvents all that is superfluous or discordant in the system of Ptolemy'. However, he did have a problem with the notion that the earth moves. As well as the familiar

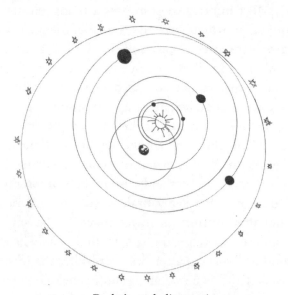

FIGURE 12: Brahe's geoheliocentric system.

objection that the 'Earth, that hulking, lazy body' is incapable of movement, he also cited the lack of any detectable solar parallax that would indicate a motion around the sun. His solution was to devise a geoheliocentric system, in which the earth remains motionless, orbited by the sun, the moon and the celestial sphere (the highest, carrying the fixed stars), while the other five planets revolve around the sun (Figure 12).

Brahe described his Tychonic system in his *On the Most Recent Phenomena of the Aetherial World*, published at Uraniborg in 1588. Thereafter, it became a competitor to both the Copernican and Ptolemaic systems.

## The mystical cosmos

By the time Johannes Kepler arrived in Tübingen, in 1589, astronomers had been struggling for two decades to incorporate Tycho's new star and supralunar comet into the prevailing Ptolemaic and Copernican systems. The publication of Tycho's new cosmology only added to their dilemma. Which model should they use for their calculations? This was not just a theoretical debate. An astronomer's duties included setting important dates such as Easter in the Christian calendar. Even more vexing was the question of which system, if any, provided a correct physical model of the heavens. Kepler's teacher at Tübingen, Michael Mästlin, had taught him that each was a mere mathematical tool with no claim on the unknowable reality of the heavens. Kepler disagreed.

Like Copernicus, Kepler believed that heliocentricity was more than a mathematical model: the earth really did move. As early as 1593, he supported a physical interpretation of the Copernican model in a student disputation at Tübingen, even going so far as to propose that the sun is the cause of planetary motions, including the earth's. It was perhaps for holding these potentially heretical opinions that Kepler was advised against pursuing a clerical career. He instead took up a post teaching mathematics at a provincial high school far to the east, in Graz, now in Austria.

It was while he was teaching in Graz that an idea struck Kepler that would haunt him for the rest of his life. He was drawing a diagram on the blackboard in front of his class when the idea came to him that the entire cosmos could be built from a concentric set of Platonic solids with the sun, rather than the earth, at its centre.

This was, of course, straight out of Neoplatonic mysticism and its dream of finding hidden messages in the heavens. The Platonic solids, so-called because they were believed to have been discovered by the Pythagoreans, are the five regular polyhedra that are constructed from only identical faces that can be fitted inside a sphere. The simplest are the cube and tetrahedron, and then the octahedron, the dodecahedron and the icosahedron. Each can be bounded by both an inner and outer sphere. Kepler realised that a set of nested spheres could be arranged so that the outer sphere of each Platonic solid forms the inner sphere of the next (Figure 13). So six concentric spheres could enclose all five Platonic solids. The number six intrigued Kepler as there were also only six (including the earth) known planets at the time. The remarkable idea that struck Kepler was that, perhaps, the six planetary spheres were arranged around the five Platonic solids.

Kepler experimented with paper models and discovered that there was only a very limited number of ways of ordering the solids so that they could be nested inside one another. Even more remarkably, if he arranged the planetary spheres in the order – Mercury – Octahedron – Venus – Icosahedron – Earth – Dodecahedron – Mars – Tetrahedron – Jupiter – Cube – Saturn, then the ratios between the size of the spheres is within about 10 per cent of the ratios between the estimated sizes of the planetary orbits in the Copernican system.

This was an extraordinary coincidence and must have been a jaw-dropping moment for the mystically minded young astronomer. Kepler was convinced that he had discovered a secret previously known only to God and those cagey Pythagoreans. He wrote up his discoveries in a booklet, *Mysterium Cosmographicum*, published in 1596, which begins with an enthusiastic declaration of faith in the wisdom of the ancients as revealed in the Copernican system.

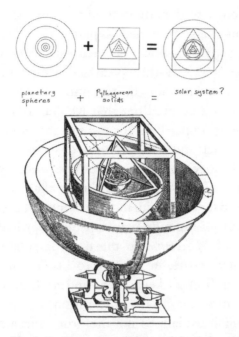

planetary spheres + Pythagorean solids = solar system?

FIGURE 13: Kepler's Platonic solids model of the solar system.
The diagram above is a schematic of Kepler's model of the solar
system illustrating the nesting of Platonic solids within the planetary
orbits. Beneath is an illustration of a model from Kepler's 1596
book *Mysterium Cosmographicum* that he had planned to be
built as a silver punch bowl.

He goes on to state his central claim that God, as a divine geometer,
had arranged the planetary spheres around this most harmonious
arrangement so that the planets can play that Pythagorean music of
the heavenly spheres.

Kepler proudly sent copies of his booklet to the leading scholars
of his day including Tycho Brahe and 'a mathematician named
Galileo Galilei, as he signs himself [who] has also been attached
for many years to the Copernican heresy'.[6] Note his almost jocular
reference to the word 'heresy', indicating how many scientists had,
by then, already absorbed Occam's insistence on the separation of
science and theology, irrespective of what the theologians taught.

The publication of his *Mysterium* not only brought the name of Johannes Kepler to the attention of the European astronomical community, but it also provided Kepler with a degree of marital capital and, in 1597, when he was twenty-six years old, he married Barbara Müller, a twenty-three-year-old widow and daughter of a successful mill-owner. Yet the reception of *Mysterium* was less of a success. Critics pointed out that Kepler could only boast an accuracy of about 90 per cent for the match between the predictions of his model and astronomical observations. He argued, quite reasonably, that the disagreement was probably due to observational errors. However, Kepler knew that to convince the sceptics he needed to obtain more accurate measurements. He asked for help from the only man capable of providing it. Kepler posted a copy of *Mysterium* to the most famous astronomer in Europe. He later wrote, in a letter to his friend and colleague Michael Mästlin, 'Let all keep silent and hark to Tycho who has devoted thirty-five years to his observations . . . For Tycho alone do I wait; he shall explain to me the order and arrangements of the orbits.'[7]

## *The silver-nosed astronomer meets the mystic dreamer*

By this time, Brahe's Uraniborg observatory had achieved international fame. A diary entry for 30 March 1590 notes that 'The king of Scotland [James VI of Scotland and the future James I of England] came this morning at the eighth hour and left at three.' However, it proved expensive to run and depended on patronage. King Frederick died in 1588 and the young King Christian was less favourably disposed towards astronomy. In January 1597, Tycho received a letter informing him that not only would the crown no longer fund the upkeep of his observatory, but it would also cease paying his state annuity. Tycho Brahe packed his instruments and left for Copenhagen. The entire Uraniborg observatory fell into disuse and eventual ruin.

After several years wandering the courts of Europe, Tycho accepted an appointment as imperial mathematician to the Holy

Roman Emperor Rudolph II. In 1598, aged fifty-two, he took up residence in Benátky Castle in Bohemia, about thirty miles outside Prague, where he set up his observatory. Later that year he received the letter from Johannes Kepler with the copy of *Mysterium Cosmographicum*.

By this time, although the Dane continued to make astronomical observations, his ambitions had shifted towards proving his geo-heliocentric model, a task beyond his mathematical skills. When *Mysterium* arrived filled with its arcane but brilliant mathematics, Brahe was sufficiently impressed to write back to offer Kepler a job.

Kepler was overjoyed to receive Brahe's letter. It had arrived in the nick of time as a religious reform was sweeping through Catholic Graz. As a strict Lutheran, the schism put his teaching post and perhaps even his own and his family's safety at risk. He immediately packed his bags and left with his family for Benátky. The two astronomers met in February 1600 when Kepler was twenty-nine and Brahe fifty-four.

Brahe immediately gave his new assistant the challenging task of making sense of the complex orbit of Mars with its twists, turns and retrograde motion. The confident young man, convinced that he had already discovered the secret of the heavens, boasted that he would solve the problem within eight days. In the event it took him eight years, yet in terms of scientific achievement they were the most productive and significant since the ancient world.

From the start, though, there were personal problems and personality clashes. The Dane had not invited the young man to his castle to nail Kepler's Pythagorean dream but to prove his own rival geoheliocentric model. Kepler's first months at Benátky were frustrating as the Danish astronomer was 'very stingy' with his observational data and would only release the bare minimum that he believed Kepler needed to pursue investigation of the Tychonic system. The pair often quarrelled, with Kepler storming off on several occasions.

Fate terminated their fiery collaboration on 13 October 1601.

Less than two years after Kepler had arrived at Benátky, Tycho Brahe was attending a banquet in Prague given by a certain Baron Rosenberg. Despite drinking heavily, the Dane felt it impolite to leave the table and endured a consequent strain on his bladder. By the time he returned home he was in great pain, and then feverish. Fever turned to delirium until, after a brief rally, the greatest naked-eye astronomer that the world has ever known succumbed and died on 24 October. Two days later, his assistant was appointed imperial mathematician and all of Brahe's painstakingly acquired data was finally within Kepler's grasp.

## Building models in the sky

In his writings, Kepler expressed sorrow at the death of his patron. Despite their social, cultural and temperamental differences, the two men nevertheless had a great deal of respect for each other. It is also not hard to discern, in Kepler's writings, his glee at having been given the keys to the Benátky Castle astronomical treasure chest. He later confessed that 'when Tycho died, I quickly took advantage of the absence . . . of the heirs, by taking the observations under my care'.[8]

In accepting the post Kepler knew that he was stepping into very large astronomical shoes. Brahe had been universally acclaimed as the greatest observational astronomer since the ancient world. If Kepler was to justify the title of imperial mathematician, then he had to equal or even surpass his predecessor's crystal-shattering discoveries. The project that he was convinced would clinch his scientific credentials was to prove that, in their five perfect solids, the ancient Pythagoreans had held the key to unlocking the heavens.

From the outset Kepler encountered problems. The difficulty he confronted is fundamental to all of science and is central to the role of Occam's razor in scientific reasoning: it is the problem of model selection. Consider the conundrum that Kepler faced. He had at least four models (Ptolemaic, Copernican, Tychonic and his own) that each accounted for the data, quite well, but not perfectly. They each gave errors of around 5 to 10 per cent. Yet there were actually a potentially infinite number of variant models that Kepler could have used as each of the above could be tweaked to improve its fit by, for example, adjusting any one of the eighty or so circles in Ptolemy's model, or by adding more entities in the form of additional epicycles to Copernicus's. With a universe of possible models to examine, where should he start?

This situation is universal in science. Remember how the scholastics spent fruitless centuries arguing where motion belonged in Aristotle's categories? In our own century, string theorists have been similarly bogged down with more mathematical models than there are particles in the entire universe. To make progress, science needs some means of sifting through the oceans of complex models that fit the data well enough to find those that are likely to yield superior models.

Many criteria may be applied to selecting a model. The most frequently adopted is dogma, whether religious, historical or cultural. Scientists, like everyone else, tend to opt for the solutions that suit their own prejudices. This was the selection criterion reluctantly adopted by Jean Buridan, and more enthusiastically by Martin Luther, to argue against a rotating earth. Copernicus had similarly allowed ancient dogma to influence his model selection when he insisted that only circular orbits were allowed in his heliocentric model. Kepler also allowed himself to be guided by his own conviction that the ancient Pythagoreans had been right. Yet, in Kepler's case, his choice of model proved fortuitous because it was simple and easy to prove wrong.

Simplicity was perhaps not foremost in Kepler's mind, but it was implied. You will remember that the Pythagorean cosmos came to him in a revelatory moment while he was teaching a class in Graz. His elation sprang from his Neoplatonist conviction that 'the heavens, the first of God's works, were laid out much more beautifully than the remaining small and common things.'[9] By 'beautiful' or, as in many other passages, 'harmony', Kepler was referring to the concept familiar to mathematicians, of mathematical beauty. The term describes the aesthetic pleasure that mathematicians enjoy through perceiving mathematical structures, whether geometric, algebraic or numerical, that possess qualities of harmony, orderliness and symmetry, but mostly of simplicity. For example, mathematicians have admired the beauty of Pythagoras's simple theorem and its elegant geometric proofs since ancient times. Four centuries after Kepler, the French mathematician Henri Poincaré wrote that 'The scientist does not study nature because it is useful to do so. He studies it because he takes pleasure in it, and he takes pleasure in it because it is beautiful . . . It is because simplicity and vastness are both beautiful that we seek by preference simple facts and vast facts.'[10] Similarly, the Nobel Prize-winning physicist Paul Dirac advised that 'The research worker, in his efforts to express the fundamental laws of Nature in mathematical form, should strive mainly for mathematical beauty.'[11] Simplicity and mathematics go hand in hand. Over the centuries, mathematicians have always striven to simplify 'ugly equations' to derive beautiful solutions. This is what mathematicians do.

Kepler later makes this even clearer in *Mysterium Cosmographicum* and *Harmonices Mundi* published in 1599, where he insisted that the world (cosmos) is a manifestation of a divine harmony, which is revealed in its construction from fundamental principles, or 'archetypes', as he calls them, 'which are indivisible in their simplicity'.[12] He goes on to declare that 'nature is simple' and frequently insists that God or the universe 'uses one cause for many effects'.[13] This is of course one of the many variants of Occam's razor that had been rumbling through the *via moderna* and one that Kepler must have encountered during his studies at Tübingen. Like Copernicus and his *via moderna* predecessors Kepler had, either explicitly in wielding the razor, or implicitly via mathematical beauty or harmony, adopted simplicity as his primary criterion for model selection.

William of Occam's compass was his determination to reduce the parts list of the world to its minimum. Neither Copernicus nor Kepler were particularly concerned with numerical simplicity but rather an aesthetic simplicity. Is this aesthetic razor the same as

Occam's? Do all roads to simplicity lead to the same destination? Even today, the jury is out on this question as simplicity is not as simple as it might appear.[14] This does not mean that simplicity is a nebulous or ephemeral concept. Many concepts in science, such as energy in physics or life in biology, are similarly slippery and hard to define; yet their slipperiness does not undermine their utility. Indeed, the indefinability of these terms hints, I believe, at their ultimate reality occupying a deeper level than our current conceptual foundations.

Kepler discovered one advantage of simple models that is rather paradoxical – that they will usually be wrong! Imagine that your friend phones you to say that she has spotted an animal in her garden and asks you to guess what kind of animal it is. You might guess 'dog' but you might also guess 'mammal'. Both are perfectly good models of the kind of animal that might inhabit a person's garden, but one is simpler. It is simpler because if we consider the 'type of animal' the dog model parameter has only one possible setting – 'dog' – whereas the mammal model has many possible settings including cat, cow, goat, dog, horse or other mammal that might conceivably be on her lawn. The simple model will be proved correct if the mystery animal barks but wrong if it meows, bleats, bahs or neighs. The more complex model will be right for all these cases but wrong if the animal chirps.

Simple models are brittle in the sense that they can be easily shattered by contrary data. In contrast, because their parameters can take on a large range of values, complex models can usually fit most data points so they are much harder to disprove. This is one of the reasons why the Ptolemaic system survived for so long: it had so many parameters that it could be fitted to almost any set of data.

Kepler experienced the brittleness of simple models when he attempted to fit his Pythagorean model to the astronomical data he had extracted from Brahe's material. No matter how hard he tried, his attempts only met with frustration. If he had been using a complex model, such as Ptolemy's or Copernicus's, then the answer would have been obvious: add more circles. With enough patience and eighty or so parameters to play with, a brilliant mathematician such as Kepler would surely have found some way of tweaking the model to fit the data. In stark contrast, there were no more Platonic solids so all Kepler could do was rearrange their ordering, yet, as we have already discovered, there are very few ways that Platonic solids may be ordered to make nested sets. Despite exhausting all of them, Kepler could not improve on his 90 per cent fit to Brahe's data.

Kepler's next step was, albeit reluctantly, to add more complexity. This is of course perfectly compatible with Occam's razor, which, contrary to many of its detractors, does not insist that the world is simple, only that in reasoning about it we should not multiply entities beyond necessity. If the existing entities cannot do the job then the razor gives you free rein to add as many entities as needed, so long as they are not 'beyond necessity'. The additional complexity that Kepler added to his model was to abandon Plato's dogma that the planets move always at uniform speed. Instead, he allowed Mars to change its velocity as it circled the sun. This additional complexity provided an immediate reward. Five epicycles in the Copernican system vanished. They became the entities beyond necessity so he removed them.

Kepler then sought to determine the radius of the perfect circle that, he believed, would describe Mars's orbit. Again he met with failure. He writes that:

> If thou [dear reader] art bored with this wearisome method of calculation take pity on me who had to go through with at least seventy repetitions of it, at a great loss of time; nor wilst thou be surprised that by now the fifth year is nearly passed since I took on Mars . . .

After five years of crunching through thousands of brain-numbing calculations (remember this was before even the slide rule had been invented) Kepler at last obtained a good fit between his predictions and four critical data points obtained from Brahe's observations. The 'hypothesis based on this method not only satisfies the four positions on which it was based, but also correctly predicts, within two minutes, all the other observations . . .' But then he laments, 'Who would have thought it possible? This hypothesis, which so closely agrees with the observed oppositions, is nevertheless false . . .'

To test his new model, Kepler plucked two additional points out of Brahe's vast store and catastrophe struck: his cherished Platonic sphere hypothesis shattered against the recalcitrant data. They now deviated from Brahe's measurements by 8 minutes of arc (the diameter of the moon is about 30 minutes of arc). Kepler lamented that: 'For, if I believed that we could ignore these 8 minutes, I would have patched up my hypothesis accordingly.' By 'patching up' Kepler meant tweaking the parameters of his model so that they gave a reasonable fit. Yet Kepler knew that his simple, and thereby brittle, model gave him very little room for manoeuvre and certainly insufficient to account for 8 minutes of arc. Kepler insisted that 'since it was not permissible to ignore them, those eight minutes point the road to a complete reformation of astronomy . . .' Kepler's only way forward was to shatter his Platonic solids and start anew.

Despite those 8 minutes of arc, Kepler suspected that he was nevertheless close to a solution. After the success of relaxing uniform motion Kepler next dared to sacrifice another ancient dogma, this time perfect circles. Nearly every astronomer since Plato had insisted that the heavenly bodies, being inhabitants of heaven, travel only in perfect circles. Of course, all circles are perfect in the sense that they are circles but by 'perfect' Plato and others were stressing perfection in the mathematical beauty sense, of an elegant, harmonious and maximally simple two-dimensional object that can, nevertheless, be described by a single number, its radius. Kepler's next, very reluctant, move was to try bending the circle. After several attempts at trying different curves he stumbled upon the ellipse, which is one of the conic sections obtained by drawing around a slice cut from a cone (Figure 14). In fact, a circle is the simplest conic section as it can be described by a single number that indicates where on the cone the (horizontal) section is cut. The next simplest is an ellipse generated by an angled cut. It requires just two numbers that identify the two points on the cone where the ellipse starts and ends. Pulled out of the cone, an ellipse is usually described as a curve drawn around two focal points, rather than the single centre of a circle. Kepler discovered that, when he bent the circular orbit of Mars into an ellipse, the predictions of his model finally fitted Brahe's meticulous observations.

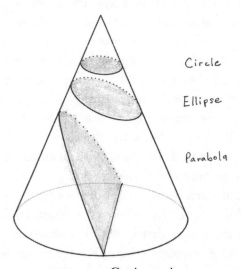

Circle

Ellipse

Parabola

FIGURE 14: Conic sections.

This was a remarkable discovery but was it peculiar to Mars? To find out, Kepler next tried adding non-uniform motion and bending the circles into ellipses for the orbits of the other planets, including the earth. To his astonishment, the predictions of his new model now matched Brahe's data perfectly. This time Kepler really had discovered the secret of the heavens.

However, the implications of his new model were staggering. For at least two millennia, the heavens had been filled with crystal spheres rotating the planets on perfectly circular orbits. Kepler's Pythagorean dream had added Plato's solids. Yet, only perfect circles can fit on the surface of a Plato's solid, the sphere. Kepler's bending of the heavenly circles had, inadvertently, shattered both its crystal and Plato's solids. Neither could fit elliptical orbits.

However, among the astronomical debris was a model of the

cosmos freed of all those cycles, epicycles and equants. It was simple. By adding just three novel steps of complexity to his simple Platonic starting model, Kepler had constructed the solar system that we know today. It remains one of the first and greatest achievements of modern science.

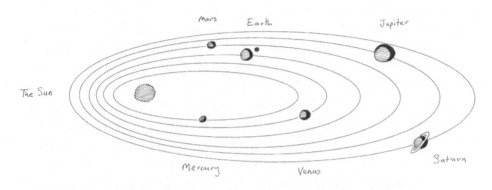

FIGURE 15: Kepler's solar system with its elliptical orbits.

However, Kepler was not proud of his discovery. He had dreamed of discovering Pythagorean harmony in the heavens but instead had found only the humble ellipse. He described it as bringing 'a cartful of dung' into the heavens.[15]

## *Laws and simplicity*

With the ancient dogmas eliminated and the spheres shattered, Kepler now glimpsed the future course of science. Having cut through the muddle of circles he glimpsed three mathematical laws that underpinned the motion of every planet in his new solar system. You will remember the value of laws from the work of the Merton Calculators whose mean-speed theorem remains in use today (though seldom correctly credited). Like the mean-speed theorem, Kepler's mathematical laws replaced arbitrary complexity with predictable laws. With laws, the world becomes simpler and thereby more predictable.

Kepler's first law states that the orbit of each planet is an ellipse with the sun at one of its two foci. His second law states that, throughout its orbit, the line from the planet to the sun sweeps out equal areas of space in equal times. So, if you draw a line from the sun out to where the planet is on its orbit at monthly intervals, you will obtain twelve sectors of the planet's ellipse. Kepler's second law insists that each sector will be of equal area. Kepler's third law states that the square of the time it takes for any planet to complete a revolution around the sun is equal to the cube of half of the length of the long axis of its elliptical orbit. This is harder to envisage but it essentially describes the relationship between the period of a planet's orbit and its distance from the sun. It is probably the most revolutionary of Kepler's three laws because it implies that distance from the sun, rather than gods, angels or any deeper philosophical principle, determines each planet's orbit. Kepler's third law made supernatural entities beyond necessity in the heavens.

As Kepler's laws are among the first known to science it is worthwhile emphasising, once again, how they made the world simpler. Before his rules, each planet was governed by its own set of rules: the size and period of its orbit and epicycles. These were arbitrary in the sense that they had to be read from the sky rather than being predicted by any more fundamental rule. Kepler's laws replaced arbitrariness with rules that governed each planet's motion. Indeed, if God had created an extra planet and placed it at a certain distance from the sun, Kepler would have been able to plot its orbit. That is the power of laws. They replace a complex, chaotic and unpredictable universe with a simple, regular and predictable cosmos.

I should point out, however, that despite dispensing with the need for supernatural entities from the heavens, Kepler believed that God had written the laws that he had managed to discern. In his magnum opus *Astronomia Nova* (*New Astronomy*), he wrote that 'Geometry is one and eternal shining in the mind of God.' For Kepler, discovering his three laws was nothing less than reading the geometrically inclined mind of God.

*Astronomia Nova*, published in 1609, describes his first two laws of planetary motion. The book was a huge success and established Kepler as the greatest astronomer of his generation. Sadly, he was not able to enjoy this new acclaim as his private life was marked by a succession of tragedies. In 1612, his wife and two sons died. Then a religious reform forced Lutherans out of Prague. Kepler had to abandon his post as imperial mathematician and move to the more tolerant Linz. He remarried but suffered from continual personal and financial troubles. Two more infant daughters died. In 1615, when he was forty-four, his mother Katharina Kepler was among fifteen women accused of witchcraft by the bailiff of Kepler's hometown of Leonberg in southern Germany. When Kepler arrived in the town, he found his mother had been chained for fourteen months to the floor of a prison cell and threatened with torture. After a trial lasting many months in which Kepler personally defended his mother, she was finally set free in the autumn of 1620, but died six months later. Eight of the other accused women were executed. Two years earlier, in 1619, Kepler had published his *Harmonices Mundi* (*The Harmony of the World*) in which he presented his third law. He had also argued that his studies had revealed harmony and simple mathematical beauty in the heavens. Sadly, the world beneath remained steeped in religious intolerance and superstition.

Kepler continued his astronomical work. Probably his most important achievement in this period was the publication in 1627 of the *Rudolphine Tables*. This monumental work consisted of an extensive star catalogue extracted from Brahe's meticulous observations, together with precise predictions of the future positions of the planets based on calculations made using his newly discovered laws. The proof of the pudding of his heliocentric system and his laws was the utility of those tables. They provided accurate predictions of planetary positions, eclipses and alignments. Their accuracy was, in the end, what most convinced astronomers of the truth of heliocentricity. Thereafter, even astrologers used Kepler's laws to predict the motions of the heavens.

At the age of fifty-eight, Johannes Kepler fell ill and died on 15 November 1630 in the German city of Regensburg. His laws remain his most lasting legacy and one of the pinnacles of scientific achieve-ment of any age. But why did they work so well? What makes the planets travel in ellipses? How do they measure their distance from the sun to know how fast to orbit? Although a huge simplification compared to the preceding cycles and epicycles, Kepler's laws remained arbitrary in the sense that Kepler had discerned them by their fit to Brahe's observational data, rather than deriving the struc-ture of the orbits from any deeper principle. Moreover, they applied only to the planets. They had nothing to say about the motion of terrestrial objects, such as arrows or cannonballs. The next great simplification was the astonishing discovery that mathematical laws not only govern heavenly motions but could also be applied on earth.

# 9

# Bringing Simplicity Down to Earth

It appears to me . . . that the matter in the heavens is of the
same kind as the matter here below. And this is because plurality
should never be posited without necessity.[1]

William of Occam, around 1323

Please observe, gentlemen, how facts which at first seem
improbable will, even on scant explanation, drop the cloak
which has hidden them and stand forth in naked and simple
beauty.

Galileo Galilei, *Dialogue Concerning the*
*Two Chief World Systems* (1632)

On 25 September 1608, a letter from the Committee of Councillors
of the Province of Zeeland was received by the States General in
The Hague, in what was then called the Dutch Republic and is
now the Netherlands, stating that an anonymous 'bearer' claimed
to have invented an instrument for seeing remote objects as if they
were near. The instrument consisted of two lenses in a sliding tube.
At its front end was a convex lens with a smaller concave lens
forming the eyepiece. The inventor requested the opportunity to
demonstrate his 'telescope' to Prince Maurice of Nassau so that he
could petition for state funds to develop the instrument further. A
week later, a spectacle maker called Hans Lipperhey from the Dutch
city of Middelburg submitted a patent application for a binocular
telescope. The following day, Jacob Metius of Alkmaar applied for
an exclusive patent for a telescopic device that he had made following

two years of research and the discovery of secret knowledge known only to him and 'some of the ancients'. Meanwhile, a 'Netherlandish inventor' was seeking to sell a working telescope at the Frankfurt Fair of 1608. A potential buyer was interested but decided that it was too expensive. By April 1609 'Dutch telescopes' were being sold in a shop on the Pont Neuf in Paris and by May the Spanish governor of Milan owned an instrument. Later that year telescopes were also circulating in Rome, Venice, Naples, Padua and London.[2]

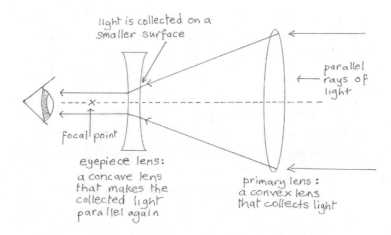

FIGURE 16: Principle of the refractive telescope.

It is often said that some ideas have their time. The ability of curved glass to magnify or distort the image of objects had been known since antiquity. Both the ancient Assyrians and Egyptians made lenses out of polished crystals and the Greeks and Romans used glass spheres filled with water to magnify objects. By the thirteenth century ground glass lenses were being used to make spectacles. Both Islamic and European scientists, such as Ibn al-Haytham (Alhazen) and Roger Bacon, experimented with the refractive properties of glass lenses. Yet, as far as we know, nobody, before the early seventeenth century, thought of combining lenses to magnify an object.

All the early accounts come from the Netherlands, making it likely that the instrument was invented there, but news of the invention spread so quickly that it is unclear who the original inventor actually was. Eventually the States General in The Hague awarded the patent to Hans Lipperhey.

The potential military advantage of an instrument that could spot distant and potentially threatening objects, such as ships and troops, was not lost on European powers almost perpetually at war. When Prince Maurice examined the new telescope in The Hague, his principal enemy, the commander-in-chief of the military forces of the Spanish Netherlands, Marquis Ambrogio Spinola, was also in attendance. By 1609 news of the invention had spread to the rest of the Spanish Empire. Archduke Albert of Austria appears to have obtained at least two Dutch telescopes in spring and winter of that year. In a letter to Cardinal Scipione Borghese, nephew of Pope Paul V, the papal nuncio in Austria, Guido Bentivoglio, described how delighted he was to peer through the archduke's telescope; and it wasn't long before a telescope was being demonstrated in Rome. Only a year or two after its invention, the telescope was still being sold as a curiosity or for its military applications. However, sometime in late spring or early summer of 1609 a young professor of mathematics with an interest in optics at the University of Padua, Galileo Galilei, was boasting that he had made his own telescope. The world was about to change.

Galileo is rightly considered to be one of the giants of science, though often for the wrong reasons. He did not prove that the earth moves, nor did he drop objects from the Leaning Tower of Pisa. However, he did make two crucially important discoveries. First, he showed that the heavens look quite a lot like the earth and are thereby likely to be governed by the same rules. Galileo's second great simplification was his demonstration that the same kind of mathematical reasoning that had proven so useful in making predictions of heavenly motions also worked on earth.

## The man who brought the heavens down to earth

Galileo Galilei (1564–1642) was born in Pisa, the eldest of six children of the musician and composer Vincenzo Galilei. In 1580, he enrolled at the University of Pisa to study for a medical degree but after attending a lecture on mathematics and acquiring his lifelong fascination with all things mathematical he shifted to natural sciences. Family financial difficulties forced him to abandon his studies before completing his degree. In the following years he attempted to establish himself as a professional mathematician, travelling between Pisa, Florence and Siena, tutoring private students or teaching at various schools. When he was only twenty-two, he published a small treatise on a new kind of weighing balance that earned him, in 1589, the position of chair of mathematics at Pisa.

Remarkably, many of his lecture notes survive from this period. Although written in his own hand, it appears that, rather than composing his own notes, he plagiarised those written by another academic, a scholar called Paulus Vallius who taught logic and the scientific method at the Collegio Romano in Rome. From these notes we can see that Galileo taught mathematics and physics in the scholastic Aristotelian tradition and was aware of the nominalist philosophers including the Merton Calculators and William of Occam, whom he refers to several times.[3]

In 1592, he obtained a position at the more prestigious University of Padua, where he lectured on mathematics, mechanics and astronomy. In 1597, he wrote to Johannes Kepler, who had only just nailed his own reputation to the mast of the Copernican system with the publication, in the previous year, of his *Mysterium Cosmographicum*. He had given two copies to a friend who travelled to Italy and one of those copies found its way to Padua and Galileo. Galileo wrote to Kepler that he had been a Copernican for several years and 'with this hypothesis [I] have been able to explain many natural phenomena which under the current hypothesis remain unexplained'. What those 'many natural phenomena' were remains a mystery.

When his father died in 1601, the thirty-seven-year-old Galileo became head of his family, responsible for supporting his younger brothers and sisters. Despite being unmarried he fathered three children with his lover Marina Gamba. His growing family brought an increasing financial burden, so as well as lecturing and private tutoring he strove to establish himself as a consultant in mathematics and the science of military engineering and fortifications.

Galileo's attentions turned to calculating the optimum number of oars for a galley and designing improved drainage pumps. He also invented an improved *sector*, the sixteenth-century equivalent of a slide-rule or calculator, used by artillery captains to calculate the optimal angle of fire for a cannon, by surveyors to measure the dimensions of a building and by merchants to calculate the value of florins in ducats, for example. Galileo's innovations brought him to the attention of the rich and powerful, including Christina of Lorraine, the wife of the Grand Duke of Tuscany, Ferdinando I, who in 1600 employed Galileo as tutor to her son Cosimo.

In May 1609, Galileo met with his friend, scholar and fellow Copernican Paolo Sarpi (1552–1623). Although a theologian, Sarpi was a sceptic, highly critical of the Catholic Church and a strong supporter of the Venetian Republic. By 1609, he had survived two assassination attempts leaving wounds that, according to Sarpi, revealed 'the style of the Roman Curia' (the Vatican court). He was also a nominalist and admirer of William of Occam, illustrating how Occam's ideas remained in circulation in the seventeenth century as part of the intellectual background of what later came to be known as the Scientific Revolution.

In a meeting with Galileo, Sarpi shared a letter received from a former student, Jacques Badovere, who described his astonishment at seeing through a magnifying eyeglass demonstrated in Paris. With only these scant details to go on, Galileo immediately returned to Padua and, within a few days, had constructed his own telescope. His first instrument magnified objects only three-fold, less than the Dutch instruments, but Galileo excelled at improving other people's inventions so it was not long before he had an instrument that could provide eight-fold magnification. Curiously, his method appears to have been entirely based on trial and error. It wasn't until 1611 that the principle of how telescopes worked was revealed by Johannes Kepler in his short book *Dioptrice*.

By 1609, Galileo had sufficiently impressed the Doge of Venice with his improved telescope for his appointment at the University of Padua to be confirmed for life and he was awarded a handsome salary of 1,000 ducats a year. With further financial backing from the Venetians, Galileo constructed a telescope that could provide thirty-fold magnification. That autumn, Galileo turned his telescope to the night sky. To do so, he introduced two innovations. The first was a mount to keep the telescope steady and the second was a circular mask fitted around the eyepiece to reduce the halo effect around bright objects seen against a dark background.

On the first night, he viewed the stars. They remained as bright pinpoints in the night sky but there were many thousands more than those visible to the naked eye. The Milky Way transformed from a pale swathe to a belt of the heavens teeming with stars. On the next night, Galileo turned his telescope to the moon. At the time all heavenly objects, including the moon, were considered to be perfectly spherical and unblemished. Galileo's first glance at the moon disproved all of that. He was astonished to see, not flawless perfection, but a rugged landscape pocked by craters and speckled with mountains. Galileo wrote that the moon's surface 'is uneven, rough, and full of cavities and prominences, being not unlike the face of the earth, relieved by chains of mountains and deep valleys'.[4] The moon was another world, not very different from the land on which his telescope was mounted.

On 7 January 1610, Galileo turned his telescope to the planets. The first peculiarity was that, unlike the stars, they no longer appeared as pinpoints of light but as bright discs hanging in space, except for Saturn that appeared to possess curious 'ears'. The planets were clearly not just wandering stars but a different kind of heavenly body. Even more remarkable was Jupiter. Galileo's telescope revealed three tiny stars that orbited neither the earth nor the sun, but around the Jovian planet. Galileo concluded 'beyond all question that there existed in the heavens three stars wandering about Jupiter as do Venus and Mercury about the sun'. Here at last was proof that, contrary to Aristotle and nearly all earlier astronomical authority, not all objects in the heavens orbit the earth.

These were stunning discoveries. The heavens were not the home of gods and angels but a realm not unlike the earth, as William of Occam had speculated nearly three hundred years earlier. Galileo wrote up his astronomical observations in his short book *Sidereus Nuncius* or *Starry Messenger*. Late in January 1610, he rushed to Venice to find a printer. Word of his cosmos-shattering discoveries had already leaked out and in February, before publication, Galileo received a letter from the secretary of the Grand Duke of Tuscany telling him that the duke was 'stupefied' by Galileo's discoveries. Galileo gambled on changing the name of his newly discovered Jovian satellites to the 'Medicean moons'. The gamble paid off. His book was published on 13 March 1610 and all 550 copies were sold out in the first week. Cosimo II de' Medici was delighted and, in May that year, Galileo moved to Florence to take up the position of the duke's mathematician and philosopher.

## But does the earth move?

When Galileo began writing *Starry Messenger*, its focus had been to describe his remarkable discoveries rather than consider their implications. Perhaps it was the arrival of the letter from Florence with its implied support from the powerful Medici family, which gave Galileo the confidence to commit to Copernican ideas. Galileo wrote of the earth that: 'For we will demonstrate that she is moveable and surpasses the moon in brightness, and that she is not the dump heap of the filth and dregs of the universe.' Note his remark about escaping from the 'filth and dregs' perspective of the medieval cosmos (see Figure 3) whose earthly centre was filled with the condemned souls of Hell. Instead, Galileo proffers a new earth as bright as the moon and as worthy as any other heavenly body.

However, it was also a world that made the theologians profoundly uneasy. Where was Hell to be located now that the sun was the centre of the cosmos? Even more importantly, where was Heaven? Medieval priests had only needed to point skyward to impress upon their congregation the proximity of God's heaven and point downwards to emphasise the importance of avoiding eternity in the fiery depths of Hell. The Church's authority was based on its claim to be mankind's guide between these supernatural realms. Yet, when Galileo's telescope found only rock in the heavens, the Church's credentials as a supernatural pathfinder were irrevocably shaken.

Perhaps to deflect controversy, *Starry Messenger* only mentions Copernicus once and Galileo offered only weak evidence in support of heliocentricity. One was his Medicean moons but, although surprising, their existence did not prove that the earth moves. His second argument was the existence of 'earth shine', reflections of the sun's rays that illuminate the dark side of the moon. However, although this was evidence that the earth was indeed a heavenly body like any other, it also did not prove that the earth moves.

Even in his magnum opus *Dialogue Concerning the Two Chief World Systems*, published two decades later, in 1632 (when he was sixty-eight years old), he offered only two additional pieces of evidence. The first was his telescopic discovery, in 1610, of phases of Venus similar to those of the moon. They only make sense if Venus orbits the sun, rather than the earth, so the observation ruled out the Ptolemaic system. However, the phases did not rule out the Tychonic geocentric system, in which an immovable earth remains at the centre of the heavens orbited by the sun, itself circled by the inner planets (Figure 12). Galileo offered one additional, but erroneous, piece of evidence, claiming that the tides are caused by the earth's motion around the sun. This was a weak argument even in the seventeenth century since it was already well known that the tides are influenced by the motion of the moon rather than the sun. It is also worth noting that, throughout his life, Galileo remained a supporter of the original Copernican system, with its many epicycles, rather than Kepler's much simpler elliptical orbit system.

The publication of Galileo's *Dialogue* in 1632 did however lead to the famous clash with the Catholic Church and retraction of his claim that the earth moves, a story that has been told many times.[5] Galileo may have been convinced by Occam's separation of science from theology, but in the eyes of the Catholic Church the Queen of Sciences remained on her throne.

## *Ironing out the lumps and bumps of the real world*

Despite Galileo's observations, in the early seventeenth century there was a stark difference between terrestrial and heavenly bodies. Whereas the motion of heavenly objects could be captured in mathematical laws, such as Kepler's, the only law of motion for terrestrial objects was the Merton Calculators' mean-speed theorem of the fourteenth century.

Galileo believed that the '[universe] is written in mathematical language'. This position was tenable in the heavens but, down on earth, objects moved mostly erratically in irregular ways that didn't appear to be governed by laws. Even the mean-speed theorem only really worked on paper. Galileo was nevertheless convinced that, despite the evidence of his senses, the motion of terrestrial objects are, like those in the heavens, governed by mathematical laws, but laws whose rules are obscured by the lumps and bumps of the terrestrial world. In an even more revolutionary move, Galileo set out to prove this hunch through experiment.

Experimentation was not entirely new. Archimedes had performed famous experiments on buoyancy and levers. The Arab physicist, astronomer and mathematician Ibn al-Haytham (965–1039) had conducted optical experiments as described in his *Book of Optics*. In his book *On the Magnet* the English philosopher William Gilbert (1544–1603) had described a range of experiments on lodestone magnets and amber, decades before Galileo. However, earlier experiments had been largely observational: observing a beam of light reflected in a mirror, or watching a lodestone attract a needle. What made Galileo's approach so revolutionary was his careful design and manipulation of the experimental environment in order to uncover the hidden regularities of terrestrial motion. It is for this reason that he is often known as the father of modern experimental science.

Around 1604, the forty-year-old Galileo initiated experiments that measured rate of fall. The problem facing him was that most objects fell too fast to be measured. He came up with an ingenious solution. Instead of allowing objects to fall fast in free space, he slowed them by rolling them down inclines fixed to tabletops. Then, to iron out the lumps and bumps of the terrestrial world, he carefully filed metal or wooden balls to make them as spherical as he could and of identical size. He cut grooves in wooden planks to ensure the balls rolled straight and lined the crevices with waxed paper to reduce friction. To measure time, he first used his pulse but then devised a more accurate dripping water clock, using a fine balance to measure the quantity of water dripped in each unit of time. He performed hundreds of experiments and averaged them to uncover regularities obscured by irregularities or *noise* in individual experiments.

His first discovery was that Aristotle was wrong. The Greek philosopher had claimed that heavy objects fall faster than lighter ones. Although there is no evidence that Galileo ever dropped objects from Pisa's Leaning Tower, he did roll light wooden balls down slopes and found they rolled at precisely the same speed as much heavier iron balls. Not only that but, rather than falling at constant speed, as Aristotle had insisted, Galileo found that they accelerated under gravity. In fact, they uniformly accelerated, obeying the mean-speed theorem discovered by the Merton Calculators and proved by the Occamist Nicole Oresme. Galileo reproduced, more or less, Oresme's graphical proof of the mean-speed theorem (Figure 7) but credited none of his medieval predecessors.

Galileo proceeded to calculate the trajectory of a projectile, such as a cannonball, by considering the ball's vertical and horizontal motions separately. He proposed that the distance travelled in the horizontal direction per unit of time would be approximately (ignoring air resistance) constant and proportional to time (0–4 seconds, in Figure 17). In the vertical direction the ball's falling motion would be uniformly accelerated according to the square of time (1–4 seconds, in Figure 17), which is a consequence of the mean-speed theorem. When he graphically combined them he obtained a parabola. This shape is interesting as, like the ellipse, it is a conic section and a hint of a link between terrestrial motion and Kepler's elliptical heavenly orbits. Yet Galileo either never read or ignored Kepler's *New Astronomy*, published nearly thirty years before his own *Discourses and Mathematical Demonstrations Relating to Two New Sciences* appeared in 1638; so as far as we know he never saw the connection.

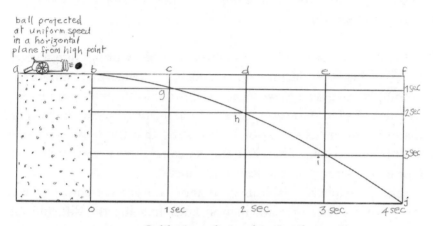

FIGURE 17: Galileo's analysis of projectile motion.

Probably the most important law discovered by Galileo is described in one of the most lucid pieces of scientific prose ever written. In his *Dialogue Concerning the Two Chief World Systems* Galileo illustrated what is known today as Galilean invariance by asking us to imagine, as Occam had, boarding a ship to illustrate the relative nature of motion.

Shut yourself up with some friend in the main cabin below decks on some large ship, and have with you there some flies, butterflies, and other small flying animals. Have a large bowl of water with some fish in it; hang up a bottle that empties drop by drop into a wide vessel beneath it. With the ship standing still, observe carefully how the little animals fly with equal speed to all sides of the cabin. The fish swim indifferently in all directions; the drops fall into the vessel beneath; and, in throwing something to your friend, you need throw it no more strongly in one direction than another, the distances being equal;

153

jumping with your feet together, you pass equal spaces in every direction. When you have observed all these things carefully . . . have the ship proceed with any speed you like, so long as the motion is uniform and not fluctuating this way and that. You will discover not the least change in all the effects named, nor could you tell from any of them whether the ship was moving or standing still.

Galileo was certainly aware of Occam, as he mentions him several times in his early lecture notes and even writes that 'motion is nothing more than a *forma fluens*' or flowing form, exactly as Occam had described it.[6] However, Galileo took the principle much further than Occam or his successors by insisting that the laws of physics are the same to an observer irrespective of their uniform motion, a principle known as Galilean invariance.

The principle of Galilean invariance is a great example of the simplifying power of mathematical laws. Imagine the difficulty of trying to calculate all the complicated motions of the objects on Galileo's ship from the perspective of the shore. Every plank, screw, nail or rope would be moving and would thereby require its own velocity. However, jump on board and, from this simpler *inertial frame*, nearly everything on the ship is still. Only the butterflies, fish, people, sails, etc. continue to move and thereby need to be provided with causes. With many thousands of motions condensed into just a handful, the world becomes simpler and easier to understand.

Like Buridan, Oresme and Copernicus before him, Galileo tossed his relativity insight into the heavens to propose that, without friction, heavenly objects could continue moving for ever. He also argued that the single daily rotation of the earth is 'much simpler and more natural' than having the sun, moon, planets and the stars rotating around the earth every day.[7] He went on to insist that his argument was 'supported by a very true maxim of Aristotle's, which teaches us that "more causes are in vain when fewer suffice"'.[8] Aristotle actually never said those words but it was a common variant of Occam's razor that had been circulating in Italy throughout the period of the *via moderna*.

In his *Discourses and Mathematical Demonstrations Relating to Two New Sciences*, Galileo's two new sciences were statics, the science of tensile strength of materials, and the science of motion, what we would call today kinematics. It described the law of inertia, the law of falling bodies and his description of the parabolic motion of projectiles. It is generally considered to be one of the most important books in the entire history of physics.

Despite, or maybe even because of, his house arrest in Arcetri after his trial, Galileo's reputation continued to grow in his final years. Several of his former students visited, including Evangelista Torricelli who would go on to invent the barometer. Torricelli arrived when Galileo was suffering fever and heart palpitations and he remained with the great scientist until he died on 8 January 1642. Another rising scientist, Robert Boyle, was also in Arcetri at that time hoping to visit Galileo. Sadly, he arrived on 9 January, a day too late.

# 10

# Atoms and Knowing Spirits

When Mr. Hobbes has recourse to what God can do (whose
omnipotence we have both great reason to acknowledge) it
imports not to the controversy about fluidity to determine what
the almighty Creator can do, but what he has actually done.

Robert Boyle, 1662[1]

In 1654, two scientists entertained a learned audience in a house
on the High Street in Oxford, not far from University College.
One of the speakers, the twenty-seven-year-old Robert Boyle (1627–
91), was a member of the assembled society known as the Invisible
College, which included other luminaries such as the mathematician
and astronomer Christopher Wren, the writer John Evelyn, and
the economist and philosopher William Petty. Boyle was tall and
handsome, with high cheekbones, straight nose and strong chin,
and spoke in a strong Irish accent though often with a pronounced
stutter. His assistant, the nineteen-year-old Robert Hooke (1635–
1703), was much shorter with a wiry hunched physique and pinched
face. Although largely unknown at this stage of his career, Hooke
would later become famous in his own right for many revolutionary
advances, such as the construction of the microscopes that he used
to discover the cellular nature of life. Both scientists were prominent
figures in what would later become known as the Enlightenment.

On this occasion, Boyle had invited the Invisible College to a
demonstration of several experiments. Most involved a vacuum
pump, which could remove nearly all the air out of a large glass
chamber. Into the chamber, Robert inserted a candle and showed

that, as the air was pumped out of the chamber, its flame flickered before extinguishing. Boyle and Hooke had demonstrated for the first time in history that fire needs air. They next inserted a watch with a loud tick. When the vessel was full of air the audience could easily hear its ticking but, as the air was evacuated, its ticking grew quieter until it could no longer be heard. The ticking returned only after Hooke allowed air to flow back into the chamber. The pair had demonstrated that sound needs air. Boyle went on to place both a magnet and a compass inside the vacuum flask and, by demonstrating that they were unaffected, showed that the force of magnetism, unlike sound, can pass through a vacuum.

Each of the above experiments were enough to amaze and startle his audience. Yet his audience was even more astonished by the pair's next demonstration. It involved a glass tube several feet long into which was inserted a lead weight and a feather. After pumping out the air, Boyle quickly inverted the glass tube. The audience watched in amazement as the lead weight and the feather fell through the length of the instrument together. Boyle had demonstrated, as Galileo had predicted, that all objects fall at precisely the same rate in a vacuum. Aristotle had indeed been wrong.

Born in Lismore, County Waterford, in Ireland into a wealthy, though self-made family, Robert Boyle did not enjoy the usual pampered lifestyle of the English or Anglo-Irish Protestant aristocracy. His father, who became the 1st Earl of Cork, had a spartan attitude to child-rearing, packing his offspring off to the country-side so that they could become accustomed to 'a coarse but clean diet and to the usual passions of the air'. The rustic life did not suit the earl's fourteenth child who was often sickly, suffering from 'ague, a defect of vision, cholic, paralysis, palsy, bloody water . . . and kidney disease'. Robert blamed these afflictions on having been forced to 'wander some wild mountains' at night in the company of an 'unskilled and drunken guide' after a fall from a horse. He also acquired a stutter such that his tutor later in life, the Frenchman Isaac Marcombes, related how 'he did stammer and stutter soe much that . . . I could scarce understand him and scarce forebear Laughing'.[2] Perhaps unsurprisingly, Robert's childhood in Ireland did not endear him to his birthplace, which he later described as a 'barbarous country'.

Aged eight, Robert was packed off to England for his early schooling at Eton College in Berkshire. The boy failed to thrive in the English public school and soon suffered from 'melancholy'.[3] He and his elder brother were sent abroad to the care of Monsieur Marcombes of Geneva. Robert appears to have been mostly happy living with the Marcombes. The family travelled widely and provided the boys with the kind of humanist education that was believed to befit an English young gentleman. This necessarily included a pilgrimage to Italy to admire those remnants of the classical civilisation that had inspired the humanists. It was on this trip that young Robert had hoped to meet his hero, the elderly Galileo, in the town of Arcetri, but arrived a day too late.

Despite the Marcombes' undoubted care of their charge, Boyle remained troubled. He relates in his memoir (which, after the fashion of the period, he wrote in the third person, adopting the name Philaretus after a Byzantine saint canonised for his unworldly generosity) how he had been plagued with religious doubts severe enough to cause him to contemplate suicide. When he was about thirteen, he was woken by a thunderstorm so fierce that 'every clap was both preceded and attended with flashes of lightning so frequent and so dazzling, that Philaretus began to imagine them the sallies of that fire that must consume the world.' Robert vowed that, if he survived the night, he would be 'more Religiously watchfully employ'd'.[4] This conflict between religious scepticism and piety was one that Boyle struggled with throughout his life. Even on his deathbed, he admitted to being continuously assailed by 'blasphemous thoughts'.[5]

By this time, England was in the throes of the Civil War. One of its earliest skirmishes was an Irish rebellion in 1641 when Robert was fourteen. He and his brother received a letter from their father with the news that he was besieged in his own castle, cutting him off from his source of wealth. The letter informed the brothers that their allowance was terminated. Being a proud man, the earl forbade the boys from returning to England in such reduced circumstances and advised instead that they either make their way back to Ireland or join the English army fighting in the Netherlands. The Marcombes supplied Robert's sturdier nineteen-year-old brother, Francis, with sufficient funds to make it back to Ireland but the more delicate Robert opted instead to return to Geneva with the Marcombes.

The conflict dragged on for several years before the seventy-five-year-old earl was forced to surrender his estate and died soon after. Robert considered that the ban on returning to England did not survive death, so in 1644, aged seventeen, after pawning jewels given to him by the Marcombes, he rode across France and bought a passage to England. From Portsmouth, he travelled to St James in London and arrived at the house where his sister Katherine lived with her four children, having been all but abandoned by her wastrel husband, the Viscount Ranelagh. On Robert's arrival she embraced her brother 'with the joy and tenderness of a most affectionate sister'[6] cementing a close relationship that they maintained throughout their lives.

While in London, Robert discovered that he was now owner of the Stalbridge estate in Dorset. The estate had an unhappy history. It had been inherited by its previous owner after informing on his father for 'unnatural practices', a seventeenth-century euphemism for same-sex relationships. After his father was hanged, the son sold the property to the earl. On the earl's death, it passed to Robert; and there he installed himself to live the life of a country squire.

The estate had, however, been devastated by the war. The manor house was ruined and most of the cottages abandoned. By cutting and selling the estate's timber, Robert managed to raise the funds he needed to hire staff who restored the house and brought the farms back into productivity. In his spare time, he wrote regularly to his sister, often including notes with simple moralising tales on all manner of topics from 'Upon the eating of Oysters', 'Upon the manner of giving meat to his dog' or 'Upon the mounting, singing and lighting of larks'. Katherine circulated his tales to her influential friends, and they proved popular enough to be published in a book of *Occasional Reflections Upon Several Subjects*. It proved so popular that Robert's style was soon lampooned by Jonathan Swift in his *Meditation Upon a Broomstick. According to the Style and Manner of the Honourable Robert Boyle's Meditations*: 'When I beheld this, I sighed, and said within myself SURELY MORTAL MAN IS A BROOMSTICK.'

Yet, despite their questionable literary value, the extra income allowed Robert to equip a laboratory at Stalbridge where he could indulge his passion for the humanist's favourite experimental science, alchemy. He wrote that 'the delights I take in it make me fancy that my laboratory is a kind of Elycium'. Although we know today that alchemy is mostly nonsense it did provide the tools to investigate the nature of matter and as such is the forerunner of the modern science of chemistry. Distillation, distinguishing acids and bases and methods for purifying metals were all first developed in alchemic laboratories. Nevertheless, alongside the sound experimental science, alchemy involved wading through a deluge of esoteric nonsense and bizarre formulae with exotic ingredients and instructions such as 'Alter and dissolve the sea and the woman between winter and spring'.[7]

The young Robert Boyle was nevertheless captivated and wrote excitedly about a worm on the 'Sombrero Coast' that transforms first 'into a tree or then into a stone'. He related the amazing tale of a 'foreign chemist' who, while travelling through France, met a tonsured monk in an inn, who claimed to have 'spirits at his command and could make them appear when he would and asked him [the chemist] if he could endure the sight of them in a terrible shape'. When the chemist remained silent the monk 'said a few words and four Wolves came into the room and run round the Table about which they were sitting for some considerable time'. They 'lookt full of rage', and 'he felt his hair stand on end, so he asked the other man to remove them, which, through a few further words, he did.' After that fright the pair enjoyed 'a banquet attended by two beautiful, well-dressed courtesans . . . Though they beckoned him, he kept his distance; he did, however, ask them questions about the philosopher's stone, "and . . . one of them writ a paper which he read and . . . understood".' Yet, as so often happens in these situations, 'both they and the paper vanished, and what was writ in it went so clearly out of Memory that he could never trace it'.

Today, all this reads like fantasy but, in the sixteenth and seventeenth century, many of the greatest intellects in Europe were struggling to make sense of these exotic ingredients, cryptic instructions and tall tales. If Robert had continued solely as an alchemist he would have remained an obscure figure in the history of an archaic 'science'. Instead he became a pivotal figure in the history of modern science. His transformation, from mystic to scientist, mirrors the emergence of modern science from its mystical and humanist roots and illustrates the value of Occam's razor in cutting through the nonsense.

## Gods, gold and atoms

By the seventeenth century, humanism was in crisis. Despairing of the mystical catacombs into which it had led, many influential philosophers were challenging the humanist trust in human creativity. René Descartes, the greatest philosopher of the seventeenth century (born a generation before Robert Boyle in 1596), had, like William of Occam, distilled contemporary philosophy to its minimalist basis insisting that, by this approach, he would 'conduct my thoughts in such order that, by commencing with objects the simplest and easiest to know, I might ascend by little and little, and, as it were, step by step, to the knowledge of the more complex'.[8] In the spirit of his famous Cartesian doubt, Descartes insisted that 'In order to seek truth, it is necessary once in the course of our life, to doubt, as far as possible, of all things.' Discarding centuries of speculation about the existence of entities beyond necessity, he arrived at just two certainties, his own existence – *cogito ergo sum* – and matter.

Like the nominalists before him, Descartes denied that the

appearance of objects corresponds to any kind of physical reality. He pointed to the observation that wax when heated changes its appearance utterly yet remains the same wax. The appearance of matter was, he concluded, an illusion of our senses. He admitted just one attribute of matter, extension, meaning the occupation of space; and insisted that only matter has extension. He famously claimed that: 'Give me extension and motion, and I will construct the Universe.' The entire universe, he insisted, was *plenum* filled only with particles.

The idea of plenum goes back to Aristotle who had proposed there was no such thing as empty space as only material objects possessed the property of extension. This seems very odd to us today but it is a good illustration of the point we explored with Kepler's abundance of models of the solar system, that there are a great, perhaps infinite, number of logically self-consistent models that may be fitted to available facts. For example, one of the facts that Aristotle cited was the observation that water will not flow out of a narrow pipe if the upper end is blocked. This observation prompted the ancient philosopher's famous dictum that 'nature abhors a vacuum'; in this case the vacuum that would be formed if the water did indeed flow out of a blocked pipe to leave only empty space behind.

Plumbing seems an odd place to start a theory of cosmic significance but the ancient philosopher also used nature's apparent abhorrence of a vacuum to dismiss one of the most prescient ideas from the ancient world, atomism. About a century or so before Aristotle was born, Democritus had claimed that matter is made up of minuscule randomly moving particles or atoms. Aristotle realised that the atomic theory contradicted his own dictum that 'everything that moves is moved by another' since there was nothing in a vacuum that could move atoms. Aristotle therefore dismissed atomism in favour of his alternative theory that matter is infinitely divisible and fills all of space, forming what was known as a plenum. He taught that the entire universe is a plenum where objects and materials, such as birds, people, arrows, fish, planets, air, water or heavenly aether, glide past each other, rather like fish sliding through water. Any gap, Aristotle insisted, would instantly be filled, just as water is sucked back into the vacuum that forms in a blocked pipe. According to this theory, empty space is a logical impossibility.

Other philosophers disagreed with Aristotle's plenum idea. The Epicureans, founded in Athens about 306 BCE by Epicurus, based their entire philosophical systems on atomism. The Roman poet Lucretius also supported atomism. The plenum vs atomism debate rumbled its way into the European medieval world where the scholastics generally sided with Aristotle. Jean Buridan found evidence for the plenum in his observation that it is 'impossible' to separate the sides of a pair of bellows if its orifice was blocked: 'not even twenty horses could do it if ten were to pull on one side and ten on the other'. William of Occam, however, leant towards an atomist position, insisting that 'both matter and form are divisible, and have parts distinct in places and position'. He even speculated that boiling and condensation were caused by rearrangement of the parts or atoms of water.[9]

The Renaissance humanists had largely abandoned Aristotle in favour of his teacher Plato, and most tended to side with the atomic theory, particularly as the interactions of atoms might provide a rational basis for understanding natural magic. For example, the alchemists claimed that atoms could be arranged in different ways to provide the elements of earth, air, fire and water, from which they thought metals such as mercury, tin and gold were composed. In an alchemist's mind the difference between base metals (those that are not gold or silver) and gold was simply a matter of rearranging their atoms, which could be accomplished by the right kind of natural magic. Hence, their dream of transmuting base metals into gold. Moreover, the followers of Paracelsus believed that the paths of the planets influenced the motion of atoms within the body to cause illness or health and that tendency could be influenced by sympathetic terrestrial objects. So if you were sad because you were suffering under the influence of a melancholy planet, such as Saturn, then you might be advised to wear a yellow gown and a gold bracelet and perhaps enjoy wine from a golden goblet as these were both sympathetic with the cheerier sun. Of course, the treatment might well work. Self-consistent, though wrong, models can work often enough to convince the credulous that they are on the right track.

The atom vs Aristotle's plenum idea rumbled on for more than two millennia and neatly illustrates how, given any set of data, it is always possible to build several, perhaps an infinite number of, self-consistent models of the universe. As we shall see, one of the principal roles of Occam's razor is to sort through rival models of our world.

Descartes accepted Aristotle's argument for the plenum yet adopted the notion that matter nevertheless came in particulate, though infinitely divisible, form. In a huge step towards modern science, he also stripped his matter particles of any humanist sympathies with the planets or magic. These became entities beyond necessity in Descartes's materialistic universe. Instead, matter consisted solely of tiny particles that moved in whirling vortexes that formed the air, water, earth, fire, plants and animals. God made the particles and gave them their first divine push; but, thereafter, their motions were entirely mechanical. Even the human body, he insisted, was 'just a statue or a machine made of earth'. Most of Descartes's philosophical ideas, including his mechanistic atomism, were described in his book *The World*, published around the time of Robert Boyle's birth, while his *Discourse on the Method* was published in 1637 and his *Principles of Philosophy* in 1644.

Although strongly opposed by Catholic humanists, Cartesian mechanistic philosophy resonated with the more nominalist, empiricist and Lutheran-inspired perspective popular in Protestant countries. By 1649, most of Descartes's works had been translated into English and were avidly consumed by the leading lights of the budding Scientific Revolution. However, many English philosophers and theologians feared that his atomist and deterministic universe was only one step away from atheism. Their fears were realised in the philosophy of Thomas Hobbes, known as 'the Monster of Malmesbury', who published his infamous *Leviathan* in 1651 when Boyle was twenty-four.

Hobbes (1588–1679) was a nominalist who took William of

Occam's reductionist approach further than anyone had previously dared.[10] He accepted Occam's unknowable God and elimination of universals and insisted that concepts such as good or evil have no philosophical or logical foundation. Like Descartes, Hobbes asserted that the universe consisted only of mechanistic particles, but he went much further than his French predecessor to eliminate the distinction between the natural and supernatural by claiming that both God and the soul are, like man, made only of matter.

For Hobbes, there was only one world. In his hugely influential *Leviathan* he argued that the only thing we can know about an omnipotent God is that He is the 'first cause of all causes', and man was merely another form of atoms in motion. He pointed out that without a benevolent God to watch over us life is naturally beset with conflict, violence and 'solitary, poore, nasty, brutish, and short'.[11] He argued, on nominalist grounds, that good and evil are merely the names we give to 'appetites and aversions'.[12] He thereby urged mankind to abandon praying to an unknowable and uncaring God but instead use human ingenuity, politics and science to build a 'commonwealth' with the aim of maintaining order, reducing suffering and increasing happiness. As the American philosopher and political scientist Michael Allen Gillespie surmises: 'Science, as Hobbes understands it, will thus make it possible for human beings to survive and thrive in the chaotic and dangerous world of the nominalist God.'[13]

Hobbes's ideas provoked consternation among conservative philosophers, including members of a group known as the Cambridge Platonists, and particularly the theologian and philosopher Henry More (1614–87). More and his Cambridge colleagues accepted the broad outline of Descartes's and Hobbes's mechanistic universe, yet insisted that mechanism alone was insufficient to account for phenomena, such as gravity, magnetism or nature's abhorrence of a vacuum. Instead, they argued for a retreat from nominalism back to Platonic realism in which an invisible 'spirit of nature' pervaded the entire universe, acting as God's agent to ensure that events turned out according to His divine plan.[14] Religion was not yet ready to release its hold on science.

## *The brave nothingness*

The end of the English Civil War heralded the return of the Irish estates to the Boyle family. Finding himself wealthy once again, in 1654, Robert, now aged twenty-seven, decided to move to the more intellectually stimulating environment of Oxford. There he built another laboratory and hired Robert Hooke.

It was around this time that Boyle tired of the arcane knowledge of alchemy and its theories that, he complained, were 'like peacock feathers, [that] makes a great show but are neither solid nor useful'. Although he still retained an interest in alchemy throughout his life, his experimental investigations shifted away from that strange 'worm on the Sombrero Coast' and its esoteric relatives to more serious science. The stimulus may have come from his sister, Katherine Jones, Viscountess Ranelagh. She was an extraordinary woman with a deep interest in science, philosophy, nature and politics and a friend of the poet John Milton and the polymath and writer Samuel Hartlib. Robert wrote to the Marcombes of how he had become acquainted with many 'men of letters' at his sister's home in London as well as members of the 'Invisible College'.

As the starting point for his science, Boyle had adopted the Cartesian idea of a thoroughly mechanistic universe but, as a devout Christian, he was appalled by Hobbes's vision of a materialist God. The solution proposed by the Cambridge Platonists was also unacceptable to Boyle, as their 'spirit of nature' smacked of paganism. Yet, rather than getting embroiled in the philosophical debate that raged around him, Boyle followed the example of his childhood hero, Galileo, in dedicating himself to executing carefully designed experiments that could settle the argument.

The object that first sparked his interest was one that went to the heart of the plenum/atomist debate. It was the humble airgun. He described in a letter how he became intrigued by an air pistol that 'could send forth a leaden bullet . . . with a force to kill a man at 25 or 30 paces' but was charged only with 'the sole impression of the air'. Airguns may seem an unlikely stimulus for a scientific revolution, but unlike the Cimmerian darkness or the philosopher's stone, they are at least real. Boyle could purchase one, take it apart, discover how it worked and demonstrate its mechanism to his sister. Most importantly, unlike the unverifiable claims of alchemy, it could be productively investigated. Hundreds of years later, the twentieth-century biologist Peter Medawar would observe that science is 'the art of the soluble'.[15] Boyle had found a soluble problem.

There were many inspirations for Boyle's experimental methodology. In his book *Novum Organum*, published in 1620, the English philosopher and statesman Francis Bacon (1561–1626) had argued, from a nominalist perspective, that the only route to scientific knowledge was to conduct lots of carefully recorded individual observations that could be tabulated and generalised to reach conclusions in the method known today as induction. Bacon was not of course the first person to use inductive arguments. For example, William of Occam made an inductive argument three centuries earlier[16] when he wrote that 'Every human being can grow, every donkey can grow, every lion, and so for other particular cases; therefore every animal can grow.' Both Occam and Bacon offered the logic of induction as an alternative to Aristotle's syllogistic reasoning, which had been undermined by the nominalist elimination of the universal.

Robert Boyle's insight was to recognise that, when combined with Galileo's carefully crafted experiments, Bacon's method of induction could become the engine for drawing sound conclusions from repeated laboratory experiments. Unlike Galileo, who only left scant details of his experiments, Robert Boyle provided extremely detailed accounts of his experimental equipment and his precise methodology, recording details, even down to the temperature or the weather, together with his raw data and analysis. For this reason Boyle, like Galileo, is a candidate for the title of father of experimental science.

Galileo had achieved fame as a physicist and astronomer. Inspired by his alchemic investigations, Boyle's interests were more down-to-earth and chemical. Instead of investigating the trajectory of the projectile fired from an airgun, Robert turned to the cause of its firing. Airguns are cocked by a piston that compresses a chamber filled only with air. Releasing the trigger removes the inward pressure, allowing the compressed air in the chamber to push the piston up the barrel thereby ejecting the bullet at some speed from the gun. The question that intrigued Boyle was how does the 'sole impression of air' push a bullet out of an airgun? Or indeed, what exactly is this invisible thing, air? These questions were mysteries in the seventeenth century but, unlike the philosopher's stone, there were answers.

Boyle's starting point was a study initiated by Galileo's student Evangelista Torricelli (1608–47). A year before he died, Galileo received reports of an intriguing problem reported by miners using mechanical pumps to remove water from flooded mineshafts. The pumps worked by retracting a piston in a cylinder that was connected to a source of water. As the piston was withdrawn, water was sucked into the chamber, in response, so it was thought, to nature's abhorrence of a vacuum that might otherwise form in the chamber. However, the miners discovered that they could never raise water more than 33 feet, no matter how hard they pulled on the pump. They asked for Galileo's help and he persuaded Torricelli to investigate.

Miners' pumps were large and cumbersome. Inspired by Galileo's approach of reproducing the essential features of the problem in a carefully controlled laboratory setting, Torricelli filled a long glass tube with water and inverted it into a dish. However, to reproduce the miners' inability to raise the column of water, the tube had to be 10 metres tall, rising above the roof of his house in Pisa. This caused considerable consternation among his neighbours who feared he was practising witchcraft, prompting Torricelli to switch from water to mercury which, being fourteen times heavier, needed only a one-metre column to obtain the same effect. In 1643, he inverted the tube of mercury, with the top end sealed into a dish of mercury (Figure 18). Expecting that abhorrence of a vacuum would prevent the mercury flowing out of the tube, the Italian scientist was astonished to observe that, contrary to nearly two thousand years of dogma, the level of mercury did indeed fall to open several inches of vacuum above the liquid surface. Hearing of Torricelli's experiments, the French mathematician, physicist and philosopher Blaise Pascal (1623–62) carried a Torricellian tube, as it was by then called, to the peak of the Puy de Dôme mountain in central France and observed that the height of the mercury column fell during the climb but rose again on the descent. Both he and Torricelli proposed that the height of the column of mercury is held up, not by the abhorrent vacuum inside the tube, but by the weight of the atmospheric air outside. This thins, and is thereby lighter, at higher altitudes, causing the mercury column to fall. They had invented the barometer.

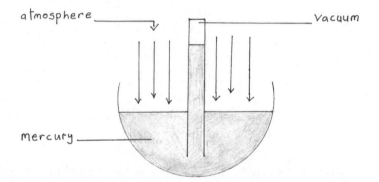

FIGURE 18: Torricelli's mercury tube experiment.

Boyle was fascinated by Torricelli's and Pascal's experiments. They inspired him to construct his own ingenious arrangements of glass and metal valves and pumps. The basic approach was to evacuate the air in a vessel large enough to do an experiment inside. He or Hooke would then insert candles, clocks, insects, fish or animals, to see how they coped with the vacuum, which brings us back to his experimental demonstrations to the Invisible College in 1654.

In his most famous experiment, Boyle pumped all the air out of a flask whose base was connected, via a valve, to a piston. When he opened the valve (so that the piston was now in contact with the vacuum), spectators gaped 'with no small wonder' as the piston was apparently sucked up by empty space, even when weighed down with 100 pounds (Figure 19). The spectators 'could not comprehend how such a weight could ascend, as it were, by itself'. The Lord Chief Justice of England, Matthew Hale, declared his admiration for Boyle's 'brave nothingness'.

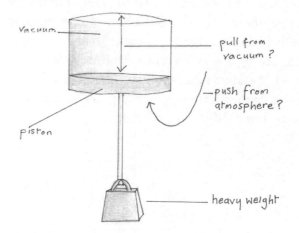

FIGURE 19: Boyle's famous demonstration of the apparent power of a vacuum to lift heavy weights.

The experiment dramatically confirmed Torricelli's and Pascal's experiments. Nature really does not abhor a vacuum after all. Indeed, in another experiment, Boyle showed that, inside an evacuated chamber, there was no resistance at all to opening up a vacuum. Like Torricelli, Boyle insisted that nature's apparent abhorrence is just the weight of the outside air. Instead of the vacuum pulling the piston into the empty chamber, the outside air was applying a push.

Boyle realised that his experiments on the nature of air were pertinent to the plenum/atomism debate. He proposed that the air's 'spring' might be provided by 'such a heap of little bodies lying upon one another, as may be resembled to a fleece of wool'. Alternatively, instead of a static model, air might consist of trillions of tiny randomly moving particles 'so whirl'd round, that each Corpuscle endeavours to beat off all others'. This was pretty much Descartes's model of 'whirling vortexes' but with vacuum, rather than plenum, between the corpuscles. The plenum, Boyle insisted, was now an entity beyond necessity.

Boyle described his vacuum tube experiments in his revolutionary book entitled *New Experiments Physico-Mechanical, Touching the Spring of the Air, and its Effects*, published in 1662. It caused a sensation. The philosopher Henry Power wrote, 'I have never read any such Tractate in all life, wherein all things are so curiously and critically handled, the Experiments so judiciously and accurately tried, and so candidly and intelligibly delivered.'[17] Like Galileo, Boyle wrote in the vernacular, in this case English, rather than the scholarly Latin preferred by most of his contemporaries. His scientific works were also free of philosophical or theological speculation that had bogged down earlier scientific writing. Moreover, unlike Galileo's terse reports of his experiments, Boyle's accounts were dense with detail and illustrated by drawings of pumps, glass flasks, valves and other pieces of apparatus, together with meticulous records of his observations. Boyle's approach was enthusiastically championed by members of the Invisible College. In 1660, the society had adopted a formal constitution and membership requiring a subscription of a shilling a week. A week later, Charles II signified his interest in the society and, in 1662, granted it a Royal Charter so that it became the Royal Society, with Robert Boyle as one of its founding members.

Not everyone was delighted with Boyle's experiments. The Cambridge Platonist Henry More was horrified. He was convinced that Boyle's 'monstrous' mechanistic science was as bad as Hobbes's godless philosophy and the door to atheism. In his *Enchiridion Metaphysicum*, published in 1671, he insisted that it was indeed the vacuum that was doing all the pulling because it was filled with 'a substance distinct from matter, that is a spirit or being incorporeal . . . a knowing principle able to move, alter and guide matter'. Instead of atoms of air, it was the ethereal hands of this 'knowing spirit' that drew the piston up into the chamber in order to close the abhorrent vacuum. More claimed that Boyle's experiments proved, not the push of a godless material mechanism, but the pull of a 'spirit of nature' that pervaded all of space.[18]

Boyle's reply to More in his *An Hydrostatical Discourse, occasioned by the Objections of the Learned Dr. Henry More* makes a powerful point that is crucial to the future course of science. He first admits that he cannot disprove the existence of More's 'knowing spirit' but goes on to insist that 'the phenomena, I strive to explicate, may be solved mechanically, that is, by the mechanical affections of matter, without recourse to nature's abhorrence of a vacuum, to substantial forms, or to other incorporeal creatures.' Boyle was here dismissing More's 'knowing spirit' or nature's 'abhorrence of a vacuum' not on the grounds that they had been disproved, but only because they were now unnecessary to account for the experimental facts. Any 'knowing spirit' was, he insisted, an entity beyond necessity and should thereby be eliminated from science: Occam's razor.

## How to identify good and excellent hypotheses

Boyle was at heart an experimentalist but his conflict with More forced him to defend his own theories, which led him to formulate criteria with which to identify good from bad ideas. These criteria are, I believe, just as significant a contribution to modern science as his experimental methods. He proposed ten key principles by which 'good and excellent hypotheses' could be separated from what he called the 'peacock theories'.[19] For reasons that will become clear, I will divide them into two groups.*

The first principles are probably the most familiar. He states that a good theory should be grounded in observations. This is essentially Bacon's method of inductive reasoning that is distinguished from the older deductive method of starting from a theory, such as all men are mortal. Boyle's support of induction did however

---

* The order in which I present the principles is not Boyle's.

bring him into conflict with the 'Monster of Malmesbury', Thomas Hobbes, who maintained that theory has priority over data. The conflict between Boyle and Hobbes is the focus of Simon Schaffer and Steven Shapin's 1985 study of the social history of science *Leviathan and the Air-Pump*.

Boyle's second and third principles are that a theory should be both logical and not self-contradictory. This is of course a fundamental characteristic of science but is not restricted to science. Plumbing is logical, non-contradictory and so on; as are the principles of hairdressing, cooking, basket weaving or philosophy.

Boyle's fourth and fifth principles insist that theories should be based on sufficient evidence and that excellent theories 'should enable us to forshow The Events that will form the basis from which welmade Tryals flow'. Boyle was here recommending that theories should make predictions ('forshow The Events') that could be checked in 'welmade Tryals', that is, in experiments. These criteria are the touchstone by which most scientific theories are usually judged to stand or fall today. As the twentieth-century physicist Richard Feynman insisted, 'it doesn't matter how beautiful, or even how simple, your theory is, if it doesn't make correct predictions, it's wrong.'

Although they are essential to it, these principles are not confined to science nor do they define it. Defendants in the dock are found innocent or guilty based on evidence. A chef tests a new recipe in 'welmade Tryals', just as a gardener or farmer will test out new seeds in their garden or field. Around 2600 BCE the ancient Egyptian architect of the pyramid at Meidum conducted a 'welmade Tryal' of his hypothesis of pyramid building only to see it disproved when the building collapsed (Figure 20). The success of later pyramid architects transformed their building hypotheses into working theories that underpinned structures like the Great Pyramid of Giza, which have survived for millennia. The principles of agriculture, metallurgy, architecture and every other foundation of modern civilisation were all similarly worked out and improved on over millennia by a combination of logic, observation, theory and countless unreported 'welmade Tryals'.

FIGURE 20: Testing hypotheses with 'welmade Tryals'.

Perhaps most importantly, none of the above principles are sufficient to guarantee scientific progress. Consider, for example, an astronomer in about the year 1600 attempting to choose between the Ptolemaic, Copernican or Tycho Brahe's models of the solar system. Each was based on logical mathematical principles or theories and each had survived many 'welmade Tryals' by making predictions that agreed reasonably well with astronomical observations. So how would you distinguish between them?

Fortunately, a tool of reason honed in the medieval world had survived through the Renaissance and Reformation to play a key role in the seventeenth century's scientific revolution: simplicity. In describing the next criterion, his sixth principle for good and excellent theories, Boyle wrote that 'a great part of the work of true philosophers has been, to reduce the true principles of things to the smallest number they can, without making them insufficient'. Boyle doesn't identify who the 'true philosophers' were, but in another passage refers 'to the generally owned rule about hypotheses, that "*entia non sunt multiplicanda absque necessitate*"'.[20] You don't need to understand Latin to recognise the edge of Occam's razor.

All of Boyle's remaining criteria, in one way or another, look for simple solutions. His seventh principle states that 'To frame a Hypothesis, one must see First, that it clearly Intelligible be.' Comprehensible theories tend to be simple; and, conversely, errors in complex theories, such as alchemy, usually become apparent when they are explained in the simplest possible terms. Descartes had made a similar point, insisting that 'a conception which springs from the light of reason alone; it is more certain than deduction itself, in that it is simpler.[21]

Boyle's eighth principle comes with its own parsimonious razor. It asserts of a good theory that 'it nought assume . . .'. The principle is essentially a restatement of the motto of the Royal Society 'Nullius in verba', which is usually translated as 'Take nobody's word for it'. Boyle was here essentially insisting that scientists should start their theories from the simplest base of only established facts, rather than dogma.

Boyle's ninth simplicity principle insists that a good theory should 'contradict no known Phenomena Of th' Universe'. Note that Boyle is not insisting that a new theory cannot contradict an established theory. He already allowed for this by his insistence that scientists should assume nothing. Rather he is referring to 'phenomena', which for him are facts, which must not be contradicted. It is not immediately obvious that this rule is another aspect of Occam's razor but it becomes apparent if we ask the question: by how many sets of laws does the universe operate? Most scientists would insist on only one, on the grounds of simplicity. However, that conviction is fairly recent. For example, Aristotle and most of his medieval followers believed that the heavenly bodies moved according to a different set of rules than those that governed terrestrial objects. Alchemists believed that magical principles operated in their laboratories that did not apply in their kitchens. Similarly, mystics, astrologers or homeopaths don't deny the laws of physics but they claim that an additional set of rules operate when they speak their spells or make their potions or predictions. Boyle's ninth principle guards against the proliferation of alternative theories that are self-consistent but contradicted by facts elsewhere in the world. Boyle is insisting on the smallest single set of rules for the entire universe: Occam's razor.

Boyle's tenth and final principle was that 'Next, of all good, the Simplest it must be: at least from all that is superfluous, free'. This was Boyle's last weapon against the 'peacock theories' of the occult in favour of the 'good and excellent theories' of science: it is Occam's razor. Boyle, like Occam, insisted that scientists should select the simplest theories that fit their data.

Boyle's simplicity criterion was absorbed into the science

promoted by the Royal Society and from there incorporated into the modern scientific method. It remains with us today, though its origin is seldom acknowledged or even realised. Ask any scientist whether they would support a complex theory that accounts for their data when a simple one does the job just as well. They may pause for a moment to think, and ask you additional questions, such as whether you mean all the data; but, so long as you reply, 'Yes, all the available data', then they will concede that they always opt for the simplest explanation that accounts for all of the data. That is what science, but no other way of reasoning about the world, does. Science has many toolboxes but only one razor.

In 1662, Boyle put his principles into action to come up with one of the first laws of modern science. Following his first principle that good and excellent hypotheses are based on sound observations, he conducted a series of experiments in which he measured the volume of gas (air) trapped beneath a column of mercury, as in Toricelli's experiments. He found that when he increased the height of the mercury column the volume of the gas shrank. After hundreds of observations, Boyle used the principle of induction to discern a law that was consistent with his tenth principle of good and excellent hypotheses, that 'the Simplest it must be: at least from all that is superfluous, free'. His gas law states that at constant temperature, the volume of a gas is inversely related to its pressure.* What could be simpler?

In the following centuries, two more gas laws were discovered. In 1787, the French balloonist Jacques Charles discovered that, at constant pressure, the volume of a gas is proportional to its temperature. Combining his law with Boyle's, he reasoned that he could make a gas less dense by heating it. On 27 August 1783, he launched a hydrogen balloon from the site where the Eiffel Tower now stands in Paris. The balloon flew across the city and into the French countryside pursued by mounted riders until it eventually landed

---

* Even more simply, it can be expressed as the equation $P_1 V_1 = P_2 V_2$ indicating that the product of pressure (P) and volume (V) of a gas at times 1 and 2 is constant: if one goes up, the other must go down.

in a field, only to be attacked by terrified peasants armed with knives and pitchforks. Two decades later, the French chemist Joseph Louis Gay-Lussac discovered the last of the three gas laws by demonstrating that, at constant volume, the pressure of a gas is directly proportional to its temperature.

Each of the gas laws is simple, even boringly simple, yet together they account for trillions of facts: from the bullets popping out of Boyle's airgun, the action of the barometer, the firing of a rifle, the lid popping off your kettle, the pressure of your car tyre, the dynamics of gas giant planets to the evolution of stars and the fate of our own sun. They also describe the behaviour of steam and thereby underpinned the steam-powered Industrial Revolution.

In 1668, at the age of forty-one, Robert Boyle left Oxford for London where he spent the rest of his life living with his beloved sister Katherine on Pall Mall. Next door but one to their house lived Nell Gwyn, the mistress of Charles II. Katherine hired Robert Hooke to build a laboratory in the rear of her property so that Robert could continue his experimental work. Katherine died on 23 December 1691. Robert followed her a week later after 'the grief for her death put him in convulsion fits which carried him off'.[22] Their house was demolished in 1850 and a bank now stands on the site. Next door is a blue plaque with the dedication 'In a house on this site lived Nell Gwynne from 1671–1687'. There is no memorial on Pall Mall to her neighbour who chased the ghosts out of the vacuum to make space for a simpler world.

Robert Boyle had helped to establish the principles of modern experimental science and extended the reach of Occam's razor into the interior of matter. His insistence on the need to apply a simplicity filter helped to establish the razor as an essential scientific tool. Yet, despite the simplifications of Copernicus, Kepler, Galileo, Boyle and others, seventeenth-century scientific understanding remained unduly complex. In particular, its two realms – the heavens and the earth – were governed by different laws. The next challenge would be to find a single set of rules for the cosmos.

# The Notion of Motion

*Three scientists walk into a (coffee) bar*

Following a meeting held at Gresham College on the north bank of the Thames on the evening of Monday, 24 January 1684, Robert Hooke, together with two of his Royal Society fellow members, Edmund Halley and Christopher Wren, made their way to a local coffee house.

This was thirty years after Hooke and Boyle had shown their revolutionary vacuum demonstrations to members of the Invisible College in Oxford. Since then, Hooke had become experimental curator to the Royal Society and had performed many ground-breaking experiments, including demonstrating capillary pressure and, after building his own microscopes, discovering the incredible diversity of microbial life. The two other scientists with him were equally eminent. The fifty-two-year-old mathematician, anatomist, astronomer and geometer Christopher Wren had, with Boyle, been one of the founding members of the Royal Society. After the Great Fire destroyed much of London in 1666, Robert Hooke was appointed surveyor and he recruited his childhood friend Christopher Wren to set about rebuilding the city, including its magnificent St Paul's Cathedral. Aged only twenty-eight, Halley was the youngest of the group, yet he had already established his reputation as one of the cleverest men in the country. As an under-graduate at Oxford, he had published papers on the moon and sunspots. In 1676, he abandoned his lectures to sail to the island of St Helena in the south Atlantic to observe an eclipse of both the sun and the moon and to catalogue stars in the southern sky.

On a day predicted by Kepler's laws, 7 November 1677, Halley made one of the first observations of a transit of Mercury across the face of the sun.

The name of the coffee house that hosted this historic meeting has, sadly, not been recorded but the Turk's Head on Exchange Alley, Joes or the Vulture are all possible venues as they were close to Gresham College and frequent haunts of Robert Hooke.[1] This was in an age before regular newspapers so, as they walked into a fog of tobacco smoke mingled with the aromas of roasted coffee, chocolate, and human sweat, they would have been greeted with cries of 'What news, sirs?' Hooke, in particular, was well known to the clientele of the London coffee houses and had even conducted experiments on their premises, including dropping a bullet from the ceiling to the floor of Galloways to demonstrate, so he claimed, the rotation of the earth. The new arrivals would have faced several rows of well-dressed men in periwigs – popular headgear since the Restoration of the monarchy in 1660 – sitting around rectangular wooden tables strewn with pamphlets, leaflets, ballads, a few candlesticks and the occasional spittoon. The throng would have been engaged in loud and lively discussions of the latest news from abroad, accounts of recent trials from the local Inns of Court, or local gossip, such as the king's recent award of a title to his bastard son by Robert Boyle's neighbour Nell Gwyn.

The company cleared a space for the distinguished new arrivals, who were likely to bring tales of the strange new science being conducted or reported at the Royal Society. However, on this occasion, the scholars eschewed the crowd to seek a quiet corner as they had business to discuss. Once seated, a boy served them fresh hot coffee for a penny a man, with unlimited refills.

His astronomical observations on St Helena had fired the young Halley with enthusiasm for astronomy and a determination to help solve the outstanding puzzles of planetary motion. First among these was the shape of the planetary orbits. Johannes Kepler, who had died fifty-four years earlier, had left to posterity his three planetary laws and elliptical orbits. Kepler's laws worked, but neither Kepler nor anyone else understood them in the sense of their having been derived from any deeper law that explained why the planets orbited the sun or followed elliptical paths. Moreover, Kepler's laws applied only in the heavens where change was very rare. On earth, change was the rule. To incorporate the terrestrial realm, science needed to incorporate the causes of change.

Halley reflected also on Galileo's principle of inertia, which explained why anything that is already moving at constant speed and direction continues to do so. However, it can only account for motion in a straight line. Twists, turns or orbits needed additional inputs. Kepler had speculated that planets were deviated from their natural tendency to move in straight lines by some 'force . . . fixed in the Sun' but hadn't elaborated his idea. Twenty-five years before the coffee house meeting, the Dutch astronomer Christiaan Huygens had provided an equation for centripetal forces that could hold bodies, at least in principle, in circular motion. Several scientists, including Halley, had noted that when combined with Kepler's third law the strength of Huygens's centrifugal force would be inversely proportional to the square of the distance of a planet from the sun; what is known today as an inverse square law. Halley wondered whether an inverse square law, applied to the sun's influence on the planets, would generate Kepler's ellipses.

Robert Hooke insisted that he already had the answer; yet when his friends probed him for details, he grew evasive, insisting that they must first attempt to solve the problem themselves so that they could come to appreciate the difficulty of the task and the ingenuity of its eventual solution. To settle the matter, Wren offered a prize of a book worth the princely sum of forty shillings to whichever of the pair could provide a convincing proof.

Time passed and nobody came forward to claim the prize. In March 1684, Edmund Halley received news that his father had vanished from his home in Islington. Five weeks later, his father's body was discovered washed up on the shore of a river east of London: he had clearly been murdered. As he had died intestate, Halley was forced to endure many months of legal wrangling that, in August 1684, took him to Alconbury, close to Cambridge. From there, Halley took the opportunity of visiting the nearby city's university to meet one of its brightest scholars who, he thought, might be able to help with Wren's challenge.

## *The lawmaker*

On 4 January 1643, a year after the young Robert Boyle had visited Italy hoping to meet the elderly Galileo, a baby was born prematurely in Woolsthorpe Manor in the county of Lincolnshire. Though sickly and small enough to fit into a quart mug, the child survived.

Isaac's early life improved little after his difficult birth. His mother, Hannah Ayscough, had been widowed three months before he was born and, three years later, she remarried and left Isaac in the care of his maternal grandmother. The boy was never reconciled to his mother or stepfather and grew up a loner with a reputation for secrecy and vindictiveness. Educated first at the King's School, Grantham, where he was taught by the Platonist Henry More, Isaac was admitted to Trinity College, Cambridge in 1661, aged nineteen. There he was befriended by the Lucasian Professor of Mathematics, Isaac Barrow (1630–77), who recognised the young man's mathematical genius. Barrow, a Neoplatonist humanist like More, also encouraged the young man's life-long interest in hermeticism and alchemy. In 1667, aged twenty-four, Newton was elected a fellow of his college and when Barrow resigned his chair in 1670 Isaac was elected as his successor. In his professorial position Newton was required to lecture at least once a week on geometry, arithmetic, astronomy, geography, optics, statics, or some other mathematical subject. Newton chose to lecture on optics, but his lectures were apparently so dull that he was often left speaking to empty seats.

After Halley and Newton had talked together for some hours, Halley asked Newton what kind of curve he thought might be described by planets held in their orbits by a solar force whose strength diminished by an inverse square law. Without hesitation, Newton replied that it would be an ellipse. Halley was dumbfounded. He asked Newton to provide a proof but, after some rummaging around in his drawers, the Cambridge scholar claimed that he could not find his notes, though he promised to post them. Halley returned to London, suspecting that Newton's proof would prove to be no more tangible than Hooke's; though the idea that an inverse square law might generate an ellipse had likely been seeded in Newton's mind by a letter he had received from Robert Hooke in December 1679.[2] Yet in November 1684 a messenger delivered a nine-page treatise to Hooke entitled 'On the Motion of Bodies in Orbit'.

Reading it, Halley was astonished to discover that Newton's treatise provided elements of an entirely new science of dynamics that not only described motion, but incorporated mathematically defined causes. He returned to Cambridge and persuaded Newton to write a book that, he assured Newton, would be published by the Royal Society. Unfortunately, by the time that Newton's book *Philosophiæ Naturalis Principia Mathematica*, known universally today simply as *Principia*, was ready, the Royal Society had exhausted its printing budget on an unsuccessful book about fishes. So, in 1687, Halley shouldered the financial burden himself of publishing the book that is, arguably, the most important in the whole history of science.

At the core of Newton's *Principia* are three mathematically defined laws of motion, which together constitute the foundation of the science of classical mechanics. His first revolutionary step was to mathematise the causes of change of motion under the umbrella term *force*. Newton's first law states that bodies that are either stationary or uniformly moving will continue doing so unless a force acts upon them. Force, in Newton's laws, is the cause of change of motion. Most importantly, it works as well for objects on earth as in the heavens.

But, what then is force? Remember that Jean Buridan's innovation was to define impetus mathematically as 'mass times velocity'. In his second law Newton similarly defined force but replaced Buridan's velocity with acceleration: the degree of change of motion. Force became 'mass times acceleration'. In this way, objects do not need forces to remain moving at the same velocity (or to stand still), consistent with Galileo's law of inertia. Force is only required for change of motion, such as the arrow leaving the bow.

It is important to note that Newton did not attempt to describe what a force actually is, only what it does. When the object's motion changes – speeds up, slows down or changes direction – then, according to the second law, a force has acted upon it. It doesn't tell us anything about the nature of that force. If, however, through some means, you know the strength of that force, then the second law allows you to rearrange Newton's equation to calculate by how much it will accelerate the object. Its acceleration will be equal to the force that has been applied divided by the object's mass.

Newton's third law of motion states that for every action there is an equal and opposite reaction. When the archer's bowstring imparts a force to the arrow to cause it to accelerate from the string, then the arrow imparts an equal and opposite force to the bow and the bowman: the recoil.

Newton's three laws of motion at last provided a mathematically defined cause of terrestrial motion, force, but the *Principia* had been prompted by Halley's question about those planetary ellipses. To answer it, Newton tossed his mechanics into the sky. Since the planets change direction – a form of acceleration – when orbiting the sun, Newton reasoned that they must be acted on by a force that, as Kepler had guessed, must emanate from the sun. Newton discovered that if that force was proportional to the product of the mass of the planet times the mass of the sun and was, as in Huygens's centrifugal force, inversely proportional to the square of the distance of the planet from the sun, then his findings matched Kepler's orbits.* Newton's universal law of gravity states that the gravitational force between two objects is equal to a constant known as the gravitational constant, G, times the product of their masses,† divided by the square of the distance between them.

---

* Well, not exactly. In *Principia* Newton proves that a planet moving in an elliptical orbit must be under the influence of a centrifugal force with magnitude inversely proportional to the square of its distance from the sun; which is actually the inverse of Halley's question.

† Mass is defined as an object's resistance to acceleration, which is independent of gravity and thereby weight. It is roughly equivalent to the quantity of matter in an object.

If Newton had stopped there then he would have been a giant of science alongside Copernicus, Kepler and Galileo. But Newton took one more revolutionary step that earned him the accolade of the greatest physicist, possibly the greatest scientist, who ever lived. Newton noted that when terrestrial objects, such as apples, fall then, as Galileo had demonstrated, they accelerate. Connecting that fact with his first law, Newton concluded that falling terrestrial objects must be acted on by a force that acts between the earth and the object. Even more remarkably, Newton discovered that he could obtain the precise trajectory of fall if he assumed that this force was equal to the same gravitation constant G that he had applied in the heavens, multiplied by the product of the two masses divided by the distance between them. His revolutionary conclusion was that gravity, the force that bends the planetary orbits, also acts on earth to cause apples to fall from trees.

Newton had finally unified motions on earth and in the heavens into a single set of laws. Throw an apple in the air and its trajectory, as described by Galileo and Newton's laws, will be a parabola. However, remember that the parabola is, like the ellipse, a conic section (Figure 14). Throw an apple with the force of a rocket and you will launch it into an elliptical orbit around the earth, like the moon. It will become a heavenly body. The earth and the heavens are just two different regions of a cosmos governed by Newton's single set of laws.

Like Boyle a couple of decades earlier, Newton, in his *Principia*, provided the principles that had guided his revolutionary science. In Rule I of his 'Rules of Reasoning in Philosophy'[3] he insists that 'We are to admit no more causes of natural things than such as are both true and sufficient to explain their appearances' – Occam's razor. In another passage he maintains that 'Nature is pleased with simplicity, and affects not the pomp of superfluous causes'. The razor had made its way from the thirteenth century, via the *via moderna*, through to Leonardo, Copernicus, Kepler, Galileo, Boyle and now to Newton to become a central plank of modern science.

There was however a cost for Newton's simple laws. He had to introduce three new entities. First, there were forces such as gravity, which, although defined mathematically, weren't really understood, any more than Buridan had understood impetus. Newton considered force as a kind of push or a pull, with the pusher and the pushed being in contact to transmit the force, rather like Aristotle's conception of motion. Yet this motion was grossly violated by gravity, which reaches from the sun across millions of miles of empty space to cause the planets to orbit. How? Newton had no idea.

## What are laws?

Which brings us back to Buridan's impetus. You will remember that previously I queried if anything would have been different if Buridan had substituted an angel for his concept of how 'impetus' is produced. We could ask the same for Newton's gravity, or force. Could an angel, rather than gravity or force, be pushing the planets or projectiles? Would anything be different if it was? In one sense, no – though it would have to be an angel who possessed a copy of Newton's *Principia* so it could ensure that it kept to the path prescribed by his laws. If the angel needed a rulebook, then why not throw away the angel and keep the rulebook? The angel, in either Buridan's or Newton's laws, is an entity beyond necessity and should be dismissed.

However, if we dispense with the angel, then we are left with the question of where the universe keeps its rulebook? Newton believed his laws were written in God's handwriting so, rather like the Christianised Forms or universals of objects, his rulebook was stored in heaven. Yet nearly four centuries before Newton, William of Occam had insisted that 'There is no order of the universe that is actually or really distinct from the existent parts of the universe'.[4] Force and gravity are generally envisaged as being 'distinct from the existing parts of the universe' as invisible entities that pull or push on objects. Occam instead insisted that terms like these that describe the relationship between physical objects, rather than the object itself, are what he called *ficta*, what we might today call fictions, or ideas. For at least one of Newton's forces, he was right.

Remember that, in Newton's laws, you multiply the masses of two objects and divide by the distance between them squared to obtain the gravitational force that pulls them together. So bigger masses will be pulled by bigger gravitational forces. But isn't that a bit odd? Galileo famously demonstrated that, despite Aristotle, objects fall at the same rate irrespective of their mass.

This conundrum is resolved in Newtonian mechanics by the action of Newton's second law to calculate the acceleration of an object due to another object's gravitational attraction by dividing the applied force by the two masses. So you first multiply by mass to obtain the force and then divide by that same mass to calculate acceleration due to that force. With mass as both multiplier and divisor, it cancels out. So mass drops out of the complete calculation of gravitational acceleration, consistent with Galileo's observation that objects fall at the same rate, irrespective of their masses.

It works, but doesn't it appear a little suspicious? Occam would surely have had something to say about it along the lines of: why put mass into the gravity equation in the first place? Yet if we leave out mass then gravity isn't behaving like a Newtonian force at all. What then is gravity? Three centuries later, another great physicist, Albert Einstein, would ponder this same question and come up with a very different understanding of gravity.

There were however two more entities in Newton's cosmos, which do not cancel each other out: absolute space and absolute time. The need for both is apparent if we ask the question: what would the concept of acceleration mean in a universe with no reference frame to measure space or time? We could imagine, for example, Galileo's ship being hit by a sudden squall that causes it to lurch so that the pots and pans in the galley below deck are thrown this way and that. The force on each object can be calculated by measuring their mass and acceleration relative to the cabin walls or each other. However, take away all the surrounding objects, including the ship, the sea and even the planet, sun, moon and stars, and imagine just a single pot accelerated in empty space. How would we know it was accelerated – and thereby subject to a force – without reference points? Remember William of Occam's argument that quantity, such as the number two, cannot be an existing thing (universal) because a pair of chairs in one room can become four chairs merely by knocking down a wall into a room furnished with another two chairs. Force, as an existing thing, similarly seems to evaporate when the walls of Galileo's imaginary ship vanish.

What then are forces? Four hundred years earlier William of Occam wrote that

> The science of nature is neither about the things that are born and die, nor about natural substances, nor about the things we see moving around . . . Properly speaking, the science of nature is about intentions of the mind that are common to such things, and that stand precisely for such things in many statements.[5]

This is an extraordinary statement for the fourteenth century but I believe that Occam is here insisting that science is all about models. So concepts, such as impetus in his day or force in Newton's, are mental constructs (intentions of the mind) that we include in our models to make predictions (statements) about the world (stand precisely for such things in many statements). This is not to deny that these words refer to entities out there in the world; but Occam is insisting that science is not about those entities but only the statements that our models allow us to make about them. The most we can hope for is that those statements are consistent with other model-inspired statements, such as those describing the result of an experiment (another statement). If the statements are consistent then we have a consistent model of the world: science. That does not mean that our model is right, only that we haven't proved it wrong. Our scientific model remains, not in the world, as Aristotle insisted, or in some mystical realm, as Plato proposed, but, according to Occam, in our head. The ultimate reality of what is *really* out there in the world will always be beyond our reach, as unknowable as Occam's omnipotent God.

We will be returning to the nature of models as they are fundamental to the role that Occam's razor plays in science. For now, we will move on to note that Newton's gravitational law is described as 'universal' in the sense that it applies to all objects in the universe. The term is not used for his mechanical laws because, as we shall see, their utility evaporates at the smallest scales of matter. However, there is an intermediate level between planets, apples and the smallest particles where Newton's laws accompanied by Occam's razor have proven to be extraordinarily useful and the basis for much of the technology that has transformed our world.

# 12

# Making Motion Work

## *The count and the canon*

On 25 January 1798, about seventy years after Newton's death, a paper by Count Rumford entitled 'An Experimental Inquiry Concerning the Source of the Heat which is Excited by Friction' was read to the Royal Society. It described an experiment in which a cannon was bored out to make a cylinder for shot and powder.

Boring a hole in solid metal is of course a task that requires a considerable quantity of Newtonian force. Rumford provided that with two draught horses. The animals were led around a yard while harnessed to a wheel so that, as they plodded a circular course, their exertions rotated the wheel at a speed of 32 revolutions per minute. The wheel was strapped around the barrel of a cannon whose sawn-off end was forced to rotate against a rigid steel drill bit submerged in a tank of water. The count, probably dressed in military breeches, waistcoat, neck-cloth, knee-length coat and sporting a tricorn hat, looked on as, to the amazement of onlookers, but not the count, the drilled cannon generated so much heat that, within two hours, the water was boiling. Rumford reports that 'it would be difficult to describe the surprise and astonishment expressed in the countenances of the bystanders, on seeing so large a quantity of cold water heated, and actually made to boil, without any fire.' The count had demonstrated that heat is related to motion.

## *What is heat?*

Count Rumford had been born Benjamin Thompson in 1753, to a modest farming family in the New World, in Woburn, Massachusetts, a small town north of Boston. It was only 120 years after the *Mayflower* had sailed to New England, yet settlements like Boston had already developed as independent economic centres. After an unsuccessful apprenticeship as a dry goods merchant, and then as a doctor, he obtained a post as a schoolteacher. At the age of nineteen, Thompson made a massive leap up the social ladder by marrying one of the wealthiest women in the colony, a thirty-two-year-old widow named Sarah Rolfe. She had inherited land and property in a town known as Rumford in New Hampshire. His marriage provided him with a living as a gentleman farmer and he was soon appointed major of the New Hampshire militia. When the Revolutionary War broke out in 1775, Thompson abandoned his wife and infant daughter to spy for the British. With the fall of Boston, he sailed for London where he managed to set himself up as a consultant responsible for recruiting and equipping the British army fighting the Revolutionary War.

While working for the crown, Thompson developed an interest in military engineering, designing his own experiments that earned him election as a fellow of the Royal Society in 1779. His work was interrupted when he was accused of spying for the French, prompting him to flee to the Continent. There he obtained a post in Munich as adviser to the Elector of Bavaria, and invented the field kitchen, portable boilers and the pressure cooker. The Elector was so pleased that he made Rumford a Count of the Holy Roman Empire.

It was during his employment as superintendent of the Munich Arsenal that Thompson performed his most famous experiment. He describes how 'Being engaged, lately, in superintending the boring of cannon . . . I was struck with the very considerable degree of heat which a brass gun acquires, in a short time, in being bored.' That drilling generates heat had, of course, been known since the first wooden sticks were rotated to make fire, yet nobody had really noticed anything strange or notable about the process. Rather like the apocryphal fall of Newton's famous apple, sometimes observation of a commonplace phenomenon by a brilliant mind can reveal deep puzzles. In this case, the puzzle was the nature of heat.

Heat was a matter of considerable debate in the eighteenth century. The ancients had associated it exclusively with fire, which the Greeks considered to be, along with earth, air and water, one of the four principal elements. However, as the Industrial Revolution geared itself up in the eighteenth century, the nature of heat and its harnessing to power steam engines became of paramount importance. A century earlier, the German chemists/alchemists Georg Ernst Stahl (1659–1734) and Johann Joachim Becher (1635–82) provided the first clue when they pointed out that a wooden log loses mass when it is burned to ash. They named the substance that leaves combustible material *phlogiston*, from the Greek *phlox* for flame, and claimed that it was the true agent of heat and combustion. They also proposed that respiration involves a kind of combustion that releases phlogiston to heat the body, which is then reabsorbed by plants and stored in wood to be released once again in burning logs to complete a kind of phlogiston eco-cycle.

This is, so far, a sound theory that fitted many facts. However, the phlogiston enthusiasts completely ignored Robert Boyle's advice on limiting their hypotheses to only the simplest 'good and excellent theories' needed to account for their findings. The German chemist J. H. Pott (1692–1777) insisted that phlogiston was 'the chief active principle in nature of all inanimate bodies', 'the basis of colours' and 'the principal agent in fermentation'.

Like all loosely defined ideas, the phlogiston theory was also able to absorb new facts. In 1774, the English scientist Joseph Priestley (1733–1804) fractionated air to obtain gas that was 'five or six times as good as common air' at promoting combustion. Priestley reasoned that his new gas must be air from which all the phlogiston had been removed, leaving a kind of sink for phlogiston released from wood or other combustible materials. He called his new gas 'dephlogisticated air' but we know it today as the element oxygen. It combines with carbon in combustible materials to make carbon dioxide: pretty much the reverse of Priestley's phlogiston explanation but neatly demonstrating how easy it is to fit any amount of data or observations to a completely wrong model of the world, so long as you have sufficient ingenuity and imagination at your disposal.

However, many chemists noticed a bigger problem for the phlogiston theory. When some metals, such as magnesium, burn they gain, rather than lose, mass. At first sight this might be seen to disprove the phlogiston model as the loss of a substance, such as the imaginary phlogiston, from a metal surely cannot add to its mass. However, the theory's advocates were not going to give up so easily. They proposed that some forms of phlogiston possess *negative weight*. As more and more metals were shown to gain mass on combustion, even when burning in Priestley's 'dephlogisticated air', proponents of the theory were forced to retreat into an even more abstract space by arguing that phlogiston was some kind of immaterial substance, vaguely analogous to one of Plato's Forms or Aristotle's universals, a kind of essence of combustion. The theory's advocates, rather like the medieval scholastics or hermetic mystics, were content to make up as many entities as it took to fit their theory to the facts.

The phlogiston theory was finally killed off on 5 September 1775, when the French chemist Antoine Lavoisier (1743–94) presented to the French Academy of Sciences his own investigations of Priestley's 'dephlogisticated air'. Lavoisier repeated Priestley's experiments with burning metals. By carefully weighing the air or oxygen before and after the metal combustion, he was able to show that its weight decreased by the same amount that the combusting metal increased its weight. Rather than any substance being released by the burning metal, the metal was instead combining with a component of the air: oxygen.

Lavoisier went on to argue that

> chemists have made phlogiston a vague principle . . . which
> consequently fits all the explanations required of it; sometimes

the principle has weight, sometimes it has not; sometimes it is free fire, sometimes it is fire combined with earth . . . It is a veritable Proteus that changes its form every instant.[1]

Note that the French chemist was not claiming to have disproved the theory of phlogiston, rather he argued that, like Robert Boyle's 'peacock theories', the theory had become so complex that it was incapable of being disproved. In contrast, his oxygen theory was simple yet it explained all the facts. Lavoisier writes that 'there is no longer need, in explaining the phenomena of combustion, of supposing that there exists an immense quantity of fixed fire [phlogiston] in all bodies which we call combustible.' He went on to argue that 'according to the principles of good logic . . . it [phlogiston] does not exist'.[2] By this time, quoting the razor itself was unnecessary as it was universally accepted as 'good logic' in science; but, in case of any doubt, he goes on to insist that we should 'not multiply entities unless necessary'.

Lavoisier's dismissal of phlogiston demonstrates how Occam's razor had, by the eighteenth century, become so ingrained into the fabric of science that, henceforth, it was almost invisible. Yet, although he had killed off one entity, phlogiston, he went on to invent another. The problem was that phlogiston enthusiasts had claimed it to be the source of both combustion and heat. Replacing phlogiston with oxygen in combustion still left him with the problem of heat: what was it? In his paper 'Réflexions sur le phlogistique', published in 1783, Lavoisier proposed that heat was some kind of 'subtle fluid' that flowed from hot to cold bodies, which he called *caloric*.

Which brings us back to Count Rumford. You will remember that the count was 'struck with the very considerable degree of heat which a brass gun acquires'. The problem was that, according to the caloric theory, heat flows from hot to cold bodies; yet, drill, bit, cannon and surrounding water all start the experiment at the same temperature. Where then was the caloric coming from? Another puzzle was that, within the course of the experiment, the heat source appeared to be inexhaustible, contradicting the caloric theory of a finite and conserved substance or subtle fluid that flowed only from hot to cold bodies.

The clue was that the whole process starts from the motion of horses. The count reasoned that their motion imparts motion to the drill bit that imparts motion to the cannon, which imparts motion to the microscopic particles of water to heat them. He insisted that 'it is these motions . . . that constitute the heat or temperature of sensible bodies'.[3] Rumford had invented the kinetic theory of heat, which claims that heat is a measure of the motion of the particles of matter. Like phlogiston, caloric became an entity beyond necessity and heat itself was understood as a measure of motion.

## Reflections on the motive power of fire

Just as the discovery of the mass gained from combustion of metals did not immediately extinguish the phlogiston theory, so Rumford's demonstration of the inexhaustible nature of caloric did not lead immediately to the demise of the caloric theory. Some scientists cited inaccuracies in Rumford's measurements while others pointed out that the count hadn't actually demonstrated that caloric was inexhaustible. Eventually, when the cannon was bored right through or the drill bit destroyed, the caloric would have been exhausted.

Another problem was that atomism, which is the basis of the dynamic theory of heat, was still resisted by many scientists who continued to believe in Aristotle's plenum theory of endlessly divisible matter. In his hugely influential book *Reflections on the Motive Power of Fire* published in 1824, twenty-six years after Rumford's cannon-boring paper, the French engineer Sadi Carnot (1796–1832) provided the fundamental mathematical framework for understanding how heat engines work by transferring heat, as caloric, from a hot to cold reservoir to found the science of thermodynamics. Just as with Ptolemy's geocentric theory or the phlogiston theory, wrong theories can, in the hands of clever scientists, nevertheless get a lot of things right.

Rumford's theory that heat was a form of motion was a great insight but it remained nebulous until about fifty years after Rumford bored his cannon. In June 1845, the English physicist James Prescott Joule (1818–89) performed more precise experiments along the lines of Rumford's and was able to demonstrate that heat is proportional to the Newtonian concept of *kinetic energy*,* which is the energy that objects possess due to their motion. About twenty-five years later, around 1870, both the Scottish physicist James Clerk Maxwell (1831–79) and Ludwig Boltzmann (1844–1906) independently fused the kinetic theory of heat and Carnot's thermodynamics with the atomic theory of matter to found the science of statistical mechanics or modern thermodynamics. They claimed that temperature is the average kinetic energy of moving atoms, equivalent to Robert Boyle's 'corpuscles' that 'so whirl'd round, that each Corpuscle endeavours to beat off all others'. As an object heats up, its atoms move faster and so have more kinetic energy and thereby a higher temperature. When it cools down, the atoms move slower and have less kinetic energy and so the temperature is lower. Temperature and motion become two sides of the same coin and another pair of previously independent phenomena – heat and motion – were reduced to one. Caloric became another entity beyond necessity and, via thermodynamics, Newton's simple laws reached down from the heavens through the terrestrial world of cannonballs and apples and into the microscopic realm of atoms in motion.

## *The application of simplicity*

Take another look at the Boyle/Hooke vacuum chamber apparatus (Figure 19) in which, in front of spectators who gaped 'with no small wonder', an apparently empty space raised a weight of 100 pounds. Does it remind you of something? Perhaps the cylinder

---

* Kinetic energy is equal to half the mass times the square of the velocity of a moving object.

of the internal combustion engine that, if it is not electric, powers your car?

Impressed by the potential of Boyle's vacuum to lift heavy loads, scientists, inventors and engineers set about harnessing its motive power. In 1679, the French Huguenot Denis Papin (1647–1713) came up with the idea of condensing steam inside a chamber to form a vacuum that pulled a piston. He had invented the single stroke engine. In 1698, the English military engineer Thomas Savery (1650–1715) took out a patent for a water pump driven by the condensation of steam inside a cylinder. A decade later the iron-monger and Baptist lay preacher Thomas Newcomen (1664–1729) designed a similar pump, 'the Miner's Friend', to remove water from flooded mines – a big problem in the emerging coal industry.

Newcomen's pump was an atmospheric engine that, just as in Boyle's experiment, relied on the weight of atmospheric air to drive the pistons. In 1764, the Scottish inventor, mechanical engineer and chemist James Watt (1736–1819) separated the power stroke cylinder from the condensing cylinder, which made the engine much more energy efficient. He also came up with the revolutionary idea of sealing the cylinders at both ends and using the heat-driven expansion of steam to push the piston out, and the coolant-driven condensation of steam to pull the piston in. He had invented the steam engine.

The first steam engines were only useful as pumps, but Watt replaced Newcomen's rocking beam with a flywheel driven from the engine's beam by a set of gears: a rotary engine. This system was soon adopted by the mill owners of industrial Britain, such as the cotton textile manufacturer Richard Arkwright, and the cotton industry shifted from water into steam-powered hyperdrive. The Cornish mining engineer Richard Trevithick (1771–1833) had the great idea of bolting a portable steam engine onto a wheeled carriage and thereby built an autonomous motor vehicle, the Puffing Devil. On Christmas Eve 1801, the Puffing Devil carried six passengers up Fore Street in Camborne and then continued on to the nearby village of Beacon. The Industrial Revolution was on the move.

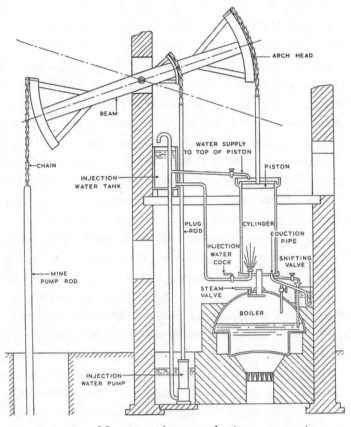

FIGURE 21: Newcomen's atmospheric steam engine.

Trevithick's engine was further improved by engineers, such as George (1781–1848) and Robert (1803–59) Stephenson, to become the powerhouse of the Industrial Revolution and the source of the biggest increase in industrial productivity that the world had ever seen. In the UK alone, coal production rose from about 20 million tons annually in 1820 to about 300 million a century later. Similarly, after centuries of stasis, crop yields rocketed as agriculture became increasingly mechanised. In the Industrial Revolution progress and productivity became exponential.* Of course, like the Renaissance

---

* Like compound interest, increase is fed back into the system to increase the rate of increase.

or Reformation, there were clearly multiple reasons for this change of gear, including the rise of capitalism and imperialism, the ready supply of coal, cheap labour, import of foreign technologies from as far away as China, bigger markets and the availability of easy profits from the slave trade. Even so, inventions such as the various kinds of steam engine undoubtedly played a huge role, yet their exponential improvement in output was only possible through the use of models, mostly drawn with ink on paper but instantiating simple laws such as those of Boyle, Newton, Carnot or Boltzmann.

Models instantiate knowledge. They model the structure, dynamics and function of machines, usually in the language of geometry and mathematics. They can be as simple as the drawing of a steam engine, such as Newcomen's in 1712 that refers to Boyle's 'weight of air' to account for the motion of the piston. The feature that makes them so useful is that improvements can be fed back into the model in a positive feedback loop that leads to exponential increases in performance. Yet models are useless without Occam's razor.

Imagine trying to build a steam engine if your model of how it works was based on More's 'knowing spirit'. How would you improve it? Perhaps you might pray to its spirit? When this approach failed, your only recourse would have been the geologically slow process of trial and error that has been responsible for most inno-vations on our planet, including life itself, since the dawn of time. That all changed when scientists and engineers shifted to using models that instantiated their knowledge and their scientific laws and, as Boyle had urged, were 'the Simplest it must be: at least from all that is superfluous, free'. With razored models in hand, scientists and engineers could predict modifications that would improve performance. If their prediction was realised, then the engineer knew that their model was good; if not, the engineer would modify their model until their predictions held true and they obtained the desired improvement. Their improved model would then be a starting point for further advances. With this positive feedback loop in place, technological development went from the linear rate of improvement delivered by trial and error to the exponential advances that characterise the modern age.

Before moving on, we should look again at Count Rumford's famous experiments. Newtonian mechanistic forces accounted for the transfer of motion and heat from the horses' labours through to the harness, the drill, the cannon and the water; however, there was a vital ingredient left out of the Newtonian chain of causes and effects: the horse. How had its labours set the whole process in motion? How had its limbs been set in motion? Can Newtonian science also account for horses and other animals, plants and microbes? Although Descartes had proposed that animals were mere machines, most eighteenth-century biologists remained sceptical and argued than the Newtonian forces were insufficient to account for life's self-propelled motions. They instead proposed that life is animated by *vital forces* that did not obey Newton's laws. To find out why, we will go fishing with horses.

# PART III

## Life's Razors

# 13

# The Vital Spark

The study of the soul is one of extreme importance. However its investigation seems to be of importance for truth as a whole and the study of nature in particular. For souls are the principle of animal life.

Aristotle, *De Anima*

The role of science is 'to substitute visible complexity for an invisible simplicity'.

Jean Baptiste Perrin

Daylight breaks over the Llanos grasslands of central Venezuela as a group of indigenous South Americans and two white Europeans set off on horseback from the village of Rastro de Abaxo in search of eels. It is the morning of 9 March 1800, the dawn of a century that will be filled with social upheaval and scientific revolutions, but Venezuela languishes in the backwaters. The local guides are probably members of one of the many Guahivo clans that then inhabited the Llanos plains lying east of the northern peaks of the Andes mountain range. Their names are unrecorded, but those of the white men are Aimé Bonpland (1773–1858) and Alexander von Humboldt (1769–1859). Bonpland is a stout, phlegmatic French botanist, while Humboldt is a slim, handsome Prussian explorer and scientist with a deep-seated interest in the nature of both life and electricity. Tragically, the native people of the Llanos have almost vanished. A wonderful account of their traditions, myths and legends was however written by one of the last Western explorers to encounter and describe them, Baron Hermann von Walde-Waldegg.[1]

The expedition is described in Humboldt's *Personal Narrative*,[2] a book that later inspired both Charles Darwin and Alfred Russel Wallace. Their guides led the white men to a stream, which, in the dry season, had shrunk to a muddy pool 'surrounded by fine trees, the clusia, the amyris, and the mimosa with fragrant flowers'. On arriving, the guides explained that the murky waters teemed with muscular eels known as *tembladores* (producers of trembling) for their painful shocks that could disable or even kill an adult. As well as being highly dangerous the *tembladores* were also notoriously hard to catch as they tended to bury themselves in the mud. Yet the locals had an ingenious method, which they described as 'embarbascar con cavallos' or 'fishing with horses'.

This phrase bewildered the explorers, but they assembled their equipment ready to dissect and study the anticipated catch, while their guides galloped away into the surrounding forest. They did not have long to wait. Before the sun had risen to its midday height, the buzz of the forest was broken by the thunder of approaching hooves – lots of hooves. The riders galloped into the clearing driving a herd of about thirty wild horses and mules. With prods from their harpoons and blows from reed sticks, the men drove the horses into the pool where their panicked commotion caused the water to boil with 'yellowish and livid eels, resembling large aquatic serpents'. The disturbed eels attacked their equine invaders by pressing themselves against the animals' bellies to deliver shock after shock that sent the horses into wild frenzies: a 'contest between animals of so different an organisation [that it] furnishes a very striking spectacle'.

Within five minutes, two of the horses had drowned. Several others had managed to stumble trembling and exhausted onto the bank, but the remainder continued to suffer repeated attacks from the eels. For a while it seemed that the equine bait would succumb to a watery grave, but the shocks gradually subsided as the fish appeared to weary and withdrew to the edge of the pool. This was when the true fishing began as the men, armed with short harpoons fastened to dry cords, were able to recover five live eels. Humboldt and Bonpland were delighted. Five shocking eels were a scientific treasure trove that would help them unravel the most hotly contested scientific debate of the eighteenth century: the nature of life.

FIGURE 22: Fishing for electric eels with horses.

Ditching centuries of mystical speculation, René Descartes had launched the mechanist revolution in the seventeenth century by insisting that all the matter in the universe, whether animate or inanimate, is composed of nothing but whirling vortexes of inanimate corpuscles. Nevertheless, mechanistic accounts of life failed to convince most scholars. Descartes compared life to the operation of a clock but few scientists took seriously the notion that clocks and cuckoos operated by the same mechanisms. The differences in vitality and complexity, already great to the naked eye, was made even more apparent through the microscope's revelation of the internal complexity of living organisms. The dominant opinion in the early nineteenth century remained that life was animated by a *vital force* that could even hurl itself out of the body of animals, such as eels, to stun other animals, such as horses. The origins of this belief lay hundreds of miles away and thousands of years previously.

## *What is life?*

Rather like Justice Potter Stewart's observation that, although he couldn't define pornography, 'I know it when I see it',[3] life is similarly easy to recognise but hard to define. Reproduction is often said to define life, yet blood or nerve cells and (most) Buddhist monks or Catholic priests fail to reproduce but are nevertheless alive. Metabolism is also argued to be one of life's defining features, but the torrent of chemical transformations that converts nutrients into waste products is hard to distinguish from the chemical reactions that cause, for example, the burning of a log. Even evolution, responsible for life's astounding variety, is dispensable for its survival, at least in the short term. If, somehow, evolution could be put on hold for, say, a million years, then, barring cataclysm, the biosphere would hardly notice.

However, most ancient people noticed that the objects around them belonged to two broad classes. The first included rocks, driftwood, sand or pebbles that are inert in the sense that they did not move unless pushed or pulled. They described these objects as *dead*. The second class comprised the crabs that crawled over the rocks, the fish that swam in the sea, the birds that flew overhead or the grasses that pushed through the dunes. These were animated in the sense of being capable of initiating their own movements. They were described as *alive*.

After having identified self-propelled motion as the signature of life, ancient philosophers next asked: what causes the motions of life that lacked an obvious mover? The answer they almost universally came up with is that living, self-propelled objects are animated by some kind of supernatural or magical *soul*. Sticking to their guns, they even classified the heavenly bodies that moved without a visible mover as living, animated by heavenly souls. Even winds, streams, storms or waves were considered animated by soul-possessing agents such as spirits, sprites, nymphs, demons or gods. As the third-century Roman chronicler Diogenes Laertius observed, 'the [ancient] world was animate and full of divinities'.[4]

## Magic, mystery and the shocking fish

The hypothesis or model that self-propelled motion = life worked pretty well. But there were anomalies. One was the lodestone, ostensibly a rock like any other but one that possessed the life-like property of animating – in the sense of moving – nails and other small iron objects. Today we know it to be a naturally magnetised form of the mineral magnetite (iron oxide), but in the ancient world lodestone was believed to be alive and possessed by a magical soul that could reach out to pull or push remote objects. As the philosopher known as Thales (born around 624 BCE in what is today Turkey) insisted, 'the lodestone has life, or soul, as it is able to move iron'. Another magical material appeared to be the yellowy-brown semi-transparent nuggets that occasionally rolled up on Mediterranean beaches that could attract fine cloth fibres or dried straw, particularly after being rubbed with a cloth. The ancient Greeks called this attractive material *electrum* but we know it today as amber.

FIGURE 23: Torpedo fish.

The ancients also knew of animals that appeared to possess the occult property of acting remotely. One was the Mediterranean

torpedo fish that, like the stingray, used its sting to deter predators, catch prey or stun hungry fishermen. Unlike the stingray,
the torpedo's sting was able to pass through a fishing line, net,
spear or trident to jolt a fisherman without even touching him.
This was widely believed to be evidence of an immaterial soul
able to reach out beyond the confines of the material body. As
the Roman poet Oppian of Corycus, writing in about 170 CE,
insisted, the torpedo fish 'exerts his magick powers and poisoned
Charms'.[5]

Magical objects and creatures were also cited as proof of
the widespread operation of magical forces in nature: the vital
principle. So Pliny the Elder (23–79 CE) argued that 'Would it
not have been quite sufficient only to cite in the instance of
the torpedo another inhabitant also of the sea, as a manifestation of the mighty powers of nature?'[6] Similarly, Alexander of
Aphrodisias (c.150–c.215 CE), who taught in Athens in around
200 CE, asked:

> Why does the magnet-stone attract only iron? Why does the
> substance called 'amber' draw up to it only chaff and dry straws,
> making them stick together? . . . No one is ignorant of the
> marine torpedo. How does it numb the body through a
> string? . . . I might prepare you a list of many such things which
> are known only by experience and are called 'unnameable prop-
> erties' by the physicians.[7]

Life, they insisted, was another kind of magic, leading to the widespread belief that sickness and health could be influenced by natural
magic.

Born in Pergamon in present-day Turkey, Claudius Galenus
(129–c.216 CE), usually known simply as Galen, was the most famous
physician in the Roman world. He had worked for several years as
surgeon to the local gladiator school, a role that provided him with
an intimate knowledge of human anatomy, before he moved to
Rome. His interest in medicine had been inspired by the earlier
theory of the Greek Hippocrates (of the famous oath) who claimed

that health depended on a delicate balance between the four humours of blood, yellow bile, black bile and phlegm. Galen added an extra spin by proposing that the four humours were kept in balance by vital spirits, or *pneuma*, that pervaded the entire universe. These vital spirits, analogous to the *qi* of Chinese medicine or the *vayu* of Indian medicine, were also thought to provide the occult forces of amber or the torpedo fish's sting. They were the material substrate of the living soul.

Galen believed that sickness is caused by an imbalance of humours and that balance could be restored by appropriate use of occult objects. So he recommended eating the flesh of the torpedo for epilepsy, whereas severe headaches should be treated by the application of a live fish to the head. Similarly, Pliny prescribed a meal made from torpedo flesh to relieve difficult childbirth, but only if the fish is 'caught while the Moon is in Libra and kept for three days in the open'. He also recommended that 'Venery [lust] is inhibited . . . if the gall of the torpedo, while it is still alive, is applied to the genitals.'[8]

Of course, even bad ideas may occasionally prove efficacious. I would guess that Pliny's prescription of placing the gall of a live torpedo on the genitals would likely blunt the desire of the lustiest youth. The cure of a live torpedo placed on the head may also have been occasionally efficacious, as electrotherapy has been shown to provide some relief for sufferers of chronic migraine.[9] However, eating the flesh of the torpedo would have provided no more benefit to the sick than any nutritious meal.

Despite his dubious potions, Galen's approach to medicine was, nevertheless, well ahead of its time. He was also a keen, though to contemporary eyes, callous animal experimenter. During one dissection of a live pig he accidentally cut the laryngeal nerve that connects the voice box to the base of the brain. He noted that its squealing immediately ceased, though its struggles continued. Galen concluded that *animal spirits*, the vital force believed to animate the living, flowed through the nerves. As far as we know, this was the first time that nerves had been associated with animal movement.

## The military advantage of magical fish

When Galen died in c.216 CE, his rational approach to medicine was largely lost, yet his magical potions and dubious treatments were transmitted faithfully through ancient texts, to the Arabs and from them to the Western world. There, they mixed with pagan and Christian magical notions to generate an occult broth of bizarre medicinal remedies.

One of the oddest was based on another magical fish, known as the *echeneis*. Pliny had described it as a 'tiny fish' that could attach to a ship and render it motionless 'with no effort of its own, neither by pushing against it nor by doing anything else but sticking to it'. An echeneis was blamed for becalming Mark Antony's flagship at the battle of Actium, rendering it an easy target for Octavian's forces. By the time the echeneis had made its way into the medieval world, it had joined the torpedo as a powerful ingredient of magical

FIGURE 24: The echeneis magical fish.

potions. Albert the Great, who you will remember taught Thomas Aquinas, described 'the ship-holder' as a fish that was sought after by magicians who used it to concoct love charms. Yet no one had ever seen an echeneis, which is not surprising as they do not exist.

The Renaissance humanists enthusiastically adopted the ancient magical objects such as the lodestone, amber and magic fish, both as medicine but also as evidence of occult forces in nature. For example, the Florentine translator of the *Corpus Hermeticum*, Marsilio Ficino, wrote that: 'The marine torpedo also suddenly benumbs the hand that touches it, even at distance through a rod'; whilst in his *Fifteenth Book of Exoteric Exercises on Subtlety*, published in 1557, Julius Caesar Scaliger (1484–1558) first high-lights the power of the torpedo that 'forces numbness into the hands' as proof of occult forces, and then goes on to criticise those who 'think they can reduce all things to fixed, manifest qualities . . .'

So to the perfectly reasonable and simple equation that autono-mous movement = life, mystics added a vast and diverse range of magical objects and properties. The trend is the same as we witnessed in other areas of science, such as astronomy (epicycles) or chemistry (phlogiston): if your model does not fit the facts then add more complexity. There were, however, a scattering of dissenters to this flight to fancy. Michel de Montaigne (1533–92), the French Renaissance humanist and contemporary of William Shakespeare, lamented that:

> How free and vague an instrument human reason is. I see ordinarily that men, when facts are put before them, are more ready to amuse themselves by inquiring into their reasons than by inquiring into their truth . . . They ordinarily begin thus: 'How does this happen?' What they should say is: 'But does it happen?' Our reason is capable of filling out a hundred other worlds and finding their principles and contexture . . . We know the foundations and causes of a thousand things that never were . . .[10]

Those 'thousand things that never were' were, of course, entities beyond necessity. Michel de Montaigne was also a nominalist and champion of the *via moderna*[11] who knew the value of Occam's razor in finding simpler models.

## *The ghost busters*

Fifty-four years before Alexander von Humboldt went fishing on the Llanos grasslands, on an April day in 1746 a group of two hundred white-robed Carthusian monks formed a line nearly two kilometres long that snaked through the grounds of their monastery in Paris. Each monk was connected to his brother by twenty-five feet of iron wire. At the end of the line stood Abbé Jean-Antoine Nollet. When all was ready the Abbé connected the leading monk to a glass bottle and, in unison with the generation of a spark, all two hundred monks jumped.

Abbé Nollet was 'Court Electrician'. This might seem a rather odd post for a mid-eighteenth-century abbot, and the Abbé was certainly not required to change prototype light bulbs or other electrical devices in the king's palaces, particularly as they hadn't yet been invented. Except for one, the Leiden jar. This had been the real star of the Parisian jumping monks show, as it provided the shock that, like that of the torpedo fish, was able to leap beyond the body (or jar). Abbé Nollet gave the jar its name after it had been invented several decades earlier in Leiden in the Netherlands as a receptacle for catching magic.

The story of Abbé Nollet's jar begins with the London physician William Gilbert (1544–1603) who noted that whereas the power of the lodestone was capable of both attracting and repelling only iron, a rubbed piece of amber was only capable of attracting diverse materials, such as fragments of chaff, wool, feathers or straw. He also discovered that if he rubbed materials such as glass, gems, ebonite, resin and sealing wax with silk or wool, then he could *charge* these materials so that they, like amber, acquired the magical power to move remote objects. He noted that these charged electrical materials would sometimes emit a spark, something that lodestone never did. He called these assorted materials *electricus* (from the Latin for amber, *electrum*) meaning 'like amber', which later gave rise to the term 'electrical'.

The discovery that the occult power of electrical objects was transferable led to the idea that it was some kind of liquid, a supernatural *subtle fluid*. It also inspired seventeenth-century mechanical enthusiasts to construct devices, known as friction machines, to capture the occult fluid. One device was a rotating sulphur globe that, when rubbed with amber or glass rods, was able to accumulate sufficient subtle fluid to attract feathers or to generate a spark when a glass rod was brought close.

Bright sparks materialising out of thin air ignited a fascination with this mysterious *electricity* and promoted its escape from the magician's laboratory into the realm of parlour tricks and circus amusements. Stephen Gray (1666–1736), a silk dyer from Canterbury, discovered that he could transfer electrical *fluid* for several hundred feet along silk threads. This inspired him to change career to become an electrical performer famous for the 'suspended charity schoolboy' trick in which a young boy was dangled from fine threads and *electrified* through his feet so that he would attract feathers and brass strips to his body. The highlight of the show was when the room was dimmed and a glass rod brought close enough to draw crackling sparks from the boy's suspended feet.

All this was fun and made for great entertainment but the mechanistic and predictable properties of *electrical fluid* gradually began to condense out of the mystical clouds. For example, Gray discovered that electricity could travel along silk threads or metal wires but not along a wooden rod, not even those claimed to be magical wands. Around 1745, Professor Pieter (Petrus) van Musschenbroek (1692–1761) from Leiden discovered that he could capture and store *electrical fluid* by transferring it from a rubbed glass rod to an insulated glass jar filled with water. One day a lawyer friend held the jar while it was charging and experienced a nasty jolt. A few days later Musschenbroek attempted the same but received such a terrifying shock that he later wrote that '[I was] hit with so much commotion that my body was rattled as if struck by a thunderbolt. My limb and entire body were terribly affected in a way that I cannot express.'

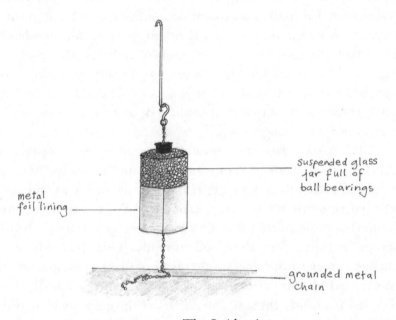

suspended glass jar full of ball bearings

metal foil lining

grounded metal chain

FIGURE 25: The Leiden jar.

Musschenbroek found that he could improve the storage capacity of his shocking device by lining it, inside and out, with metallic foil. Also, rather than water, he filled its interior with lead shot connected to a brass rod that pierced through its stopper. This was the famous electrical machine that Abbé Nollet named the Leiden jar and used to electrify his chain of monks. Its ease of manufacture, together with the discovery that several Leiden jars could be connected in series, allowed the construction of even more powerful *batteries* (named after an ammunition battery) capable of stunning lines of oxen, hefty wrestlers, soldiers and, of course, monks. With its domestication and ready availability, it wasn't long before one magic-show spectator was reminded of a far more powerful spark kindled out of the atmosphere/nature.

## Catching lightning in a jar

During a visit to Boston in 1743, a thirty-seven-year-old newspaper publisher and printer named Benjamin Franklin (1706–90) witnessed the Edinburgh *electrician* Adam Spencer demonstrate 'Sparks of fire Emitted from Face and Hands of a Boy Suspended Horizontally'. Although later to become famous as one of America's greatest statesmen, the sparking-boy trick ignited in Franklin a lifelong fascination with electricity and a determination to discover its secrets.

After several years of experimentation in his own home, in 1750 Franklin wrote a letter to the Royal Society of London, proposing that electrical objects either possessed an excess (charged positive) or were depleted (negatively charged) of the electrical fluid, or were neutral. Any imbalance, such as in the suspended boy trick, would lead, Franklin proposed, to the flow of electricity and sometimes a spark. His even more astonishing proposal was that lightning bolts are also a form of electricity, essentially gigantic sparks caused by an electrical imbalance between the clouds and the ground.

Franklin set out to prove his theory on a stormy night in a field in Philadelphia in June 1752, when he and his son flew a kite attached to a string via a key, with the other end of the string tied to the terminal of a Leiden jar. The pair pulled on the string to guide the kite into a storm cloud hoping to lure the electricity of a lightning bolt into their jar. They were fortunate in being unsuccessful as, if they had succeeded, both father and son would have been incinerated and the course of American history would probably have been very different. However, despite his inability to catch lightning, Franklin 'observed some loose threads of the hempen string [of the kite] to stand erect and to avoid one another, just as if they had been suspended on a common conductor'. It appeared that the electricity of the cloud had sufficiently electrified the kite and trailing string for its charged fibres to repel, exactly as they would have done if charged by a Leiden jar. Franklin had shown that lightning, for millennia thought to be a missile of the gods, was just another form of electricity. The German philosopher Immanuel Kant called Franklin 'the new Prometheus', and the English chemist Joseph Priestley described the kite experiment as 'the greatest, perhaps, that has been made in the whole compass of philosophy, since the time of Sir Isaac Newton', insisting that Franklin was 'the father of modern electricity'.[12]

## Is life electric?

A few years earlier, in 1746, Robert Turner, an English naturalist steeped in the vitalist tradition, had published *Electricology: Or a Discourse upon Electricity. Being an Enquiry into the Nature, Causes, Properties and Effects thereof, upon the Principles of the Aether*. The book gave a rambling account of various occult theories but departed from tradition when the author proposed that the torpedo's shock was electrical. For Turner, this did not make life any less magical because he believed electricity was a form of magic, but it did suggest that animal magic might also be captured in a Leiden jar.

Colonel John Walsh (1726–95), an English soldier, scientist and diplomat posted to India, performed a more extensive study of *animal electricity*, as it came to be known. After being elected a member of the Royal Society in 1772, he was introduced to Benjamin Franklin and the two men came up with a plan which Franklin wrote up as 'Direction to Discover whether the Power that gives the Shock in touching the Torpedo . . . is Electrical or Not'. Later that year, Walsh travelled to La Rochelle in France and enlisted local fishermen to obtain specimens of Mediterranean torpedoes. He first persuaded the men to volunteer to be shocked by the Leiden jar. They reported that 'the Effect was precisely the same as that of the Torpedo'. A public demonstration of his experiments, with perhaps a nod to Abbé Nollet, demonstrated that the torpedo shock, like Leiden jar electricity, could be passed along a human chain. He wrote to Franklin reporting that 'The Effect of the Torpedo appears to be absolutely Electrical' and an example of *animal electricity*.

Walsh's experiments had pretty much proved that the torpedo's shock was a form of electricity, but could electricity play a more fundamental role as the vital force that animated all living things, as Robert Turner had proposed? At 8.30 p.m. on the evening of 26 April 1786, Luigi Galvani (1737–98), an anatomist and physician from the University of Bologna, went into the garden of the Palazzo Poggi carrying a collection of dissected frogs' legs and spinal cords attached to brass hooks. He hung the hooks from the iron railing that surrounded the garden and, rather like Benjamin Franklin more than thirty years earlier, waited for an approaching storm. When it arrived, Galvani was delighted to witness the eerie spectacle of dismembered frogs' legs seemingly returning to life, contracting and twitching on the wall of the Zamboni garden.

Galvani's experiment was of course the inspiration for Mary Shelley's classic gothic novel *Frankenstein* but had been inspired by a chance observation, about a decade earlier, when Galvani was dissecting frogs while his assistant was hand-cranking a machine that, coincidentally, generated electric sparks. As the assistant reached for a knife close to where a frog was being dissected, a spark jumped from the blade to the animal's sciatic nerve. Galvani and his assistant were astonished to see the frog's leg twitch. Galvani poked the leg with his scalpel to check that it was dead; but it failed to budge. Galen had proposed that nerves were the conduits for vital spirits. Galvani now wondered whether those vital spirits were electrical.

## The Prussian electrician

Alexander von Humboldt was born in Berlin in 1769 to a wealthy Prussian family. When he was nine his father died, leaving Alexander and his elder brother, Wilhelm, in the care of their emotionally distant and domineering mother, Maria Elisabeth.

As a child, Alexander was fascinated by nature, collecting and studying small animals, shells, plants, fossils or rocks; so much so that he became known as 'the little apothecary'. As he grew older, he became interested in the big scientific question of whether the vital life force was spiritual or mechanical. His dream was to study natural science, but his mother's ambitions for both her sons were more prosaic: she expected them to take up respectable careers in the Prussian Civil Service. Alexander was packed off to Hamburg to study business, which he detested. He eventually persuaded his mother to allow him to follow his geological interests and study at the Freiberg Academy for the just about respectable profession of mining engineer.

FIGURE 26: Portrait of Alexander von Humboldt.

Alexander quickly advanced through his profession. He toured mines along the Rhine and wrote a book on their geology and another on the strange moulds and sponge-like plants that he discovered lurking in their damp, dark crevices. He took a sympathetic interest in the living and working conditions of miners and invented both a mask and a lamp to improve their safety. He also wrote a geology textbook for miners, and set up a school for miners' children.

During a visit to Vienna in the autumn of 1792, Humboldt

learned about Galvani's experiments and became fascinated by the possibility that life's vital force is electricity. He replicated Galvani's frog muscle work and experimented with Leiden jars to electrocute and dissect frogs, lizards and insects. He made incisions on his own arm, rubbed them with acids and poked them with metal and electrified wires. In one of his more reckless experiments he placed a zinc electrode into his mouth and a silver one into his rectum. When he connected both with a wire, he suffered abdominal pain and reported that 'by inserting the silver more deeply into the rectum a bright light appears before both eyes'.[13]

Fortunately, Humboldt survived the rigours of self-experimentation and, in 1794, visited his brother, Wilhelm, who was living with his wife, Caroline, in the town of Jena, then the cultural heart of the duchy of Saxe-Weimar and close to the home of that colossus of German culture Johann Wolfgang von Goethe. Wilhelm and Caroline were members of Goethe's circle of friends and so they provided Alexander with an introduction to the great poet.

By this time, Goethe was no longer the Adonis-like figure who had won and broken many hearts in his youth. He was fat, middle-aged and sullen. However, the arrival of the young Prussian, bursting with fascination for all things natural, helped to rekindle Goethe's youthful enthusiasms, particularly for the natural sciences. The two spent hours discussing the controversies of their day, including the clash between the vitalists and mechanists over the nature of life. They performed experiments together, dissecting frogs and observing how their legs twitched when poked with wires.[14] They even inspected the corpses of a couple who had been struck by lightning. Goethe's romanticism, essentially a revival of Christian humanism but with a love of nature replacing its adulation of humanity, had a lasting influence on Humboldt's science and philosophy of life.

In 1790, Humboldt travelled to London and there met Joseph Banks, the botanist who had served on Captain Cook's voyages of discovery in the South Pacific. Banks's tales of discovery, together with his collections of plants and animals, instilled in Humboldt a determination to become an explorer. Yet his ambitions remained thwarted by his mother's prosaic expectations of her son, until 1796, when she died of cancer. Neither brother attended their mother's funeral. Within a month, Alexander had resigned from his position as mining inspector and begun his new career as naturalist, geographer, geologist and explorer.

In 1799 he obtained permission from the Spanish crown to explore 'Spanish America'. On 5 June 1799, he and the French botanist Aimé Bonpland set sail on the *Pizarro* for Latin America, landing at Cumaná, Venezuela, on 16 July carrying an armoury of scientific instruments including several barometers. They spent many months exploring the coastal region before heading inland to discover whether, as was rumoured, the Orinoco river joined the Amazon. After several gruelling weeks crossing the vast monotonous flatlands and 'burning plains' of the Llanos, they reached the small town of Calabozo. There they met an unexpected soulmate who had built 'an electrical machine with large plates, electrophori [an apparatus for generating static electricity], batteries, and electrometers; an apparatus nearly as complete as our first scientific men in Europe possess'. The devices had been constructed by Mr Carlos del Pozo, 'a worthy and ingenious man', who had managed to put them together from descriptions, mostly provided by Benjamin Franklin's memoirs. Señor del Pozo was extraordinarily pleased to meet Humboldt and Bonpland, particularly as they had brought with them some of the most sophisticated electrical instruments available on the planet. Indeed, he 'could not contain his joy, on seeing for the first time instruments which he had not made, and which appeared to be copied from his own'.

Their purpose in arriving in Calabozo had not however been to meet an electrician, however delightful, but to find electric fish. The locals' horse-fishing method described at the beginning of this chapter landed the European explorers five live electrical eels, or *gymnoti*, but not without incident. Humboldt describes accidentally stepping on a live eel whose shock delivered 'pain and numbness . . .

so violent . . . that I was affected the rest of the day by a violent pain in the knees, and in almost any joint'.

Humboldt and Bonpland confirmed that, like electricity, the fishes' shock passed through metal but not sealing wax. It could also pass through the bodies of both Bonpland and Humboldt when they held hands. More interestingly, Humboldt discovered that the fish were able to control and direct their shock. For example, when one of the pair held the animal's head and the other its tail, only one of them usually received a shock, which could be delivered at either end. These experiments convinced Humboldt that animal electricity was essentially the same as 'the electrical current of a conductor charged by a Leyden vial, or Volta's pile' but one that the animal could control.

Humboldt spent four more years travelling in Latin America culminating in his legendary ascent up the mighty Chimborazo mountain* in the Andes where he and Bonpland conducted the first-ever systematic bio-geography study, documenting the mountain's plant life from the tropical rainforests at its base to the lichens clinging to its rocky pinnacle. He sent regular reports that were published in European journals and shipped thousands of specimens of plants and animals, many new to science, back to Berlin or to Joseph Banks in London, ensuring that, by the time he returned to Europe, he was the most famous scientist of his age.

It wasn't until 1808 that Humboldt published his observations of the electrical eel and by then the debate had moved on from the question of whether the shocks from the fish were electrical to the more general role, if any, of animal electricity. More significant in the end was the founding, in 1811, of the university known today as the Humboldt University of Berlin by Alexander and his brother Wilhelm. In 1836, the university recruited Emil du Bois-Reymond (1818–96), a brilliant young physician and physiologist. Du Bois-Reymond designed an instrument called a galvanometer sensitive enough to detect the very weakest electrical signals travelling along nerves. In the kind of theatrical public demonstration that would

---

* Actually an extinct volcano.

have impressed Stephen Gray, he showed that he could make his galvanometer needle jump merely by contracting his arm.[15] Galen's *animal spirits*, the vital force that he had believed travelled along the nerves to provide animal locomotion, were finally shown to be the same force that provided the shocking power of torpedoes, the attractive properties of amber and the destructive force of lightning bolts. Another entity, the vital spirit, was shown to be an entity beyond necessity, at least as the agent of animal locomotion.

## The body electric

If there is anything with a claim to being life's vital spirit, it is electricity because nearly every aspect of life is dependent on electricity in some form. As well as transmitting nerve signals and animating muscles, electricity plays a vital role inside every living cell. Electrical forces fold biomolecules into the particular shapes needed to form proteins, enzymes, cell membranes, DNA, sugars or fats, and they drive all the molecular machinery involved in cell replication, locomotion, repair, photosynthesis, metabolism, sight, hearing, taste or smell. Signals travel down nerves as waves of electrically charged particles flowing in and out of the nerve cells. Nanoscale electrical turbine engines sitting in the membranes of cellular internal organs called *mitochondria* generate the energy that powers all our cells. Bacteria communicate via electrical signals that pass down nanowires[16] and bioelectric signals guide the development of embryos.[17]

Humboldt died in 1835, only a year before du Bois-Reymond was able to measure and prove the connection between electricity and nerves. Humboldt's last work was twenty-seven years in the making, a massive five-volume book, *Cosmos*, whose index ran to a thousand pages. The book was a sweeping, rambling and often brilliant attempt to bring together geography, anthropology, biology, geology, astronomy, chemistry and physics. No similar synthesis of human knowledge had been attempted since Aristotle. In it, Humboldt urges us to 'strive after a knowledge of the laws and the principles of unity that pervade the vital forces of the universe'. The vital force is still there but, in Humboldt's final intellectual synthesis, it is drained of mysticism and his book instead looks forward to the kind of physical unification/synthesis that we will be exploring later on.

While many scientists accepted that 'the laws and principles of unity that pervade the vital forces of the universe' might indeed account for the mechanisms of life, no one had any idea how its vast diversity and complexity could be accounted for by those same 'laws and principles'. Even the most hardened mechanists were unable to come up with any kind of mechanistic account of the origin of even single species, let alone the many thousands discovered each year by naturalists like Humboldt. As the American poet Joyce Kilmer (1886–1918) later lamented:

> *Poems are made by fools like me*
> *But only God can make a tree.*

Undermining this claim was the next major challenge for the science of simplicity.

# 14

## Life's Vital Direction

The theory [natural selection] itself is exceedingly simple, and the facts on which it rests – though excessively numerous individually, and coextensive with the entire organic world – yet come under a few simple and easily understood classes.

Alfred Russel Wallace (1889)[1]

The necessity of nature brings it about that the parts in some animals are conveniently arranged for the health of the whole. For example, the front teeth are sharp and apt for dividing food and the molars are flat and apt for mashing food . . . Consequently, these parts do not exist because of such uses. Rather, when they come to be then the animals survive. The reason is this . . . these parts become apt for conserving the animal by chance.

William of Occam, c.1320[2]

On 18 June 1858, a letter arrived at Down House, about half a mile outside the village of Downe in Kent. It was addressed to the eminent forty-nine-year-old naturalist Charles Darwin. Darwin's fame was largely due to the enormous popularity of his book, now most widely known as *The Voyage of the Beagle*, published nineteen years earlier. In his book, Darwin described not only his voyage on the famous ship but the astonishing variety of plants and animals that he encountered during his five-year expedition to the South Atlantic, Pacific and Indian Ocean. The feature that had both impressed and baffled the young naturalist was how each island that he visited was populated by its own peculiar cluster of distinct species. He wrote that:

> My attention was first thoroughly aroused, by comparing together . . . the mocking-thrushes, when, to my astonishment, I discovered that all those from Charles Island belonged to one species (*Mimus trifasciatus*); all from Albemarle Island to *M. parvulus*; and all from James and Chatham Islands . . . belonged to *M. melanotis*.

Why did nearby islands have their own distinctive species?

Of course, creationists had a ready answer: God chose to make the world that way. Yet, by the nineteenth century, many biologists were increasingly dissatisfied with explanations that invoked the deity. Nearly two centuries earlier, Newton had argued that 'like a watchmaker, God was forced to intervene in the universe and tinker with the mechanism from time to time to ensure that it continued operating in good working order'. Yet tinkering with every thrush or finch inhabiting every tiny island in an archipelago surely smacked of obsession?

Charles Darwin had been pondering the puzzle of the origin of species ever since his return from his voyage on the *Beagle*. He had even drafted a 'sketch' of his own theory sixteen years earlier. Yet Darwin had not published any of his thoughts as he felt he first needed to acquire more evidence. Consequently, in the preceding two decades, he occupied himself studying worms or shoreline creatures, such as barnacles, or examining specimens that he obtained from his extensive network of field naturalists. These 'fly-men', as they were commonly called, scoured the forests, jungles, swamps, savannah and deserts of the world to find and preserve the most exotic and rarest animals and plants, which they then packed off and sold to museums and wealthy naturalists.

The letter that arrived at Down House in June 1858 was from one of those fly-men, Alfred Russel Wallace. The name was known to Darwin as, a few years earlier, Wallace had written to his London agent, Samuel Stevens, describing his latest shipment, noting that

'The domestic duck var. is for Mr. Darwin.'[3] Moreover, in 1855, and rather unusually for a fly-man, Wallace had written a scientific paper entitled 'On the Law which has regulated the Introduction of New Species'.[4] Wallace had even written to Darwin asking his opinion on the theory described in his paper but had not received a reply. Yet the paper and subsequent letter must surely have alerted Darwin to the fact that this hardly known fly-man was also pondering the origin of species.

The letter that arrived on Darwin's doorstep in 1858 was different from the last in that it enclosed a draft manuscript. When Darwin started reading it, he was thunderstruck. It began by citing Malthus's 1798 'Essay on the Principle of Population', which pointed out that reproduction routinely outstrips available resources. Wallace went on to argue that 'the life of wild animals is a struggle for existence' so that only a small fraction of those that are born manage to survive to reproduce. He continued to argue that 'those that die must be the weakest . . . while those that prolong their existence can only be the most perfect in health and vigour.' Wallace discussed how domestic animal breeders had performed a kind of artificial selection of desirable characteristics, such as docility or plumpness, to turn wolves into domestic dogs, or wild boar into pigs. Finally, he argued that 'the struggle for existence' similarly acts on the natural variation in wild species so that, Wallace noted, 'the weakest and least perfectly organized must always succumb'. This process, taking place over millennia, led, Wallace argued, to evolutionary change and the establishment of new species each adapted to their local environment.

Wallace had solved the mystery of the origin of species. He closed his letter by requesting that if Darwin found any merit in his (Wallace's) paper, then he pass it on to the most eminent geologist in England, and Darwin's close friend, Charles Lyell.

It is easy to imagine Charles Darwin's beetle-brow creasing and his (clean-shaven at this time) jaw dropping as he leafed through the pages of Wallace's manuscript. When he had recovered, he wrote to his friend Lyell, passing on Wallace's paper with a letter admitting that:

Your words have come true with a vengeance that I should be forestalled . . . I never saw a more striking coincidence. If Wallace had my M.S. sketch written out in 1842 he could not have made a better short abstract! Even his terms now stand as Heads of my Chapters . . . So all my originality, whatever it may amount to, will be smashed.

He continued that 'I hope you will approve of Wallace's sketch, that I may tell him what you say'. He also promised to write to Wallace and pass his paper on to a scientific journal.

## Butterflies and beetles

Alfred Russel Wallace had been born in 1823, one of nine surviving children. His mother, Mary Anne, came from a well-to-do Hertford family. However, according to Alfred, his father 'lived quite idly' before embarking on a series of disastrous business ventures that lost most of the family's fortune. By 1816, the downwardly mobile family were forced to move from their large house in London to cheaper accommodation in Monmouthshire on the Welsh border, where Alfred was born.

When Alfred was five, the family's prospects brightened after the death of a relative brought an inheritance that allowed the family to move to their mother's hometown of Hertford. However, another disastrous business venture sent the family's fortunes tumbling once again and the Wallaces had to make use of their only growing resource, their children. As soon as each of Alfred's elder brothers was of age, he was apprenticed, in turn to a surveyor, a carpenter and then a trunk-maker. The family were forced to downsize to a series of smaller houses until they moved to premises too small to house all of the children. In desperation, Alfred was sent to board in a private school where he gave tuition to the younger boys to pay for his own fees.

However, the family's worsening finances forced the termination of his formal education when Alfred was aged fourteen. He was sent to lodge with his elder brother, John, then serving an apprenticeship to a building firm in London and where Alfred earned sixpence a day as a general odd-jobs boy. Happily, London provided Alfred with plenty of free opportunities to further his self-education. He visited the British Library, the Zoological Gardens as well as the Hall of Science (now Birkbeck College) on Tottenham Court Road, one of seven hundred Mechanics Institutes that had been set up by wealthy philanthropists to promote science among working people. It was there where Alfred first heard, and later met, the Welsh socialist and co-founder of the Cooperative Movement, Richard Owen, whose utopian socialism and scepticism towards established religions would play a big role in shaping Alfred's own ideas. He later wrote that 'the only beneficial religion was that which inculcated the service of humanity, and whose only dogma was the brotherhood of man'.

FIGURE 27: Butterfly specimen board.

In 1837, Alfred began an apprenticeship as a surveyor and spent the next six years travelling up and down the country, often dealing with claims in relation to the General Enclosure Act. This had impoverished many peasant farmers by fencing off common land previously held for grazing. Alfred considered it to be 'legalized robbery of the poor'. However, the work did entail lots of countryside walks, helping to kindle Alfred's lifelong interest in zoology, ornithology, botany and entomology, particularly of beetles.

When his father died in 1843, Alfred, aged twenty, was forced to abandon his surveying apprenticeship to take on any building work he could find. After many months spent as a jobbing labourer, he eventually found a position more suited to his interests, as a teacher in Leicester. In his spare time he visited the city's local library, where he read Alexander von Humboldt's *Personal Narrative*, Charles Darwin's *The Voyage of the Beagle* and Thomas Malthus's *An Essay on the Principle of Population*. He also met his lifelong friend Henry Walter Bates (1825–92), another self-educated young man who had also acquired an interest in beetles. Alfred and Henry made regular excursions into the Leicestershire countryside to return with catch nets full of beetles, butterflies and other insects. The pair would then carefully mount each specimen on a wooden board nailed to the wall inside the Bates's garden shed. The next challenge was to label each specimen with a species name. The pair meticulously noted features, such as wing colour, markings and size and, crucially, learnt how to distinguish between the natural variation of these characteristics within a species and the variation that separates species. It was this exercise that sparked Alfred's interest in the key question of nineteenth-century biology: how are species made?

## Sticks, stones and the origin of species

Most Victorians, who thought about the question at all, believed that all species on earth had been created within one week about six thousand years before. So, to the average Victorian, the diversity of plants and animals was no mystery. Like heavenly motions at the time of William of Occam, the natural world was accounted for by the existence of a God who had made the cattle, beasts and the creeping things 'after their kind' for man 'to have dominion over'. He, after all, was the only entity powerful enough to fill the world with such an immense number and variety of plants, animals and creeping things.

In many ways, the biblical account of creation was the last vestige of Aquinas's science of theology that William of Occam had taken his razor to six hundred years earlier. Aquinas had interpreted the biblical creation story within Aristotle's philosophy, equating Aristotle's final cause of cats, dogs and oak trees, with their assumed role in God's plan. He had also insisted that God had equipped each species with its own universal of *dogness, catness, oakness* and so on; so that species were doubly immutable. Occam's banishment of both final causes and universals undermined this philosophical realist claim for species immutability. Moreover, as can be seen from the quotation opening this chapter, Occam also considered that natural variation in characteristics, such as teeth, might have arisen by chance and be retained because 'the animals survive'. This was a remarkable premonition of the theory of natural selection but one that, like so many others, was buried in the Enlightenment's loathing for anything medieval. Dogma prevailed so, by the eighteenth century, the Swedish father of modern taxonomy, Carl Linnaeus (1707–78), insisted that 'There is no such thing as a new species.'[5]

The need for a creationist origin of species was famously bolstered by the watchmaker argument of the English clergyman, naturalist and philosopher William Paley (1743–1805). Paley argued that the mechanistic laws described by Newton, Galileo, Boyle or Faraday were incapable of generating the level of organised complexity of, say, the human eye. To illustrate his point, he imagined that, when walking across a heath, he stumbled upon 'a watch upon the ground, and it should be inquired how the watch happened to be in that place'. He went on to insist that: 'There must have existed, at some time, and at some place or other, an artificer or artificers, who formed

[the watch] for the purpose which we find it actually to answer . . .'
Paley's *intelligent design* argument invoked what is sometimes termed
the 'god of the gaps', a divine explanation for a phenomenon that
cannot be accounted for by the known natural laws.

However, by the early eighteenth century new discoveries were
driving the god of the gaps into a very tight corner. For example,
alongside the watch that William Paley imagined, he might have
spotted rocks on his heath that looked remarkably like living plants
and animals. These *figured stones*, as they were sometimes called,
were regularly tossed up by farmers' ploughs or found on beaches.
Some resembled the branches of trees, others leaves, seeds or fragments of bone. More puzzling were those that looked like unknown
sea creatures. Dorset farmers regularly turned up giant disc-shaped
stones whose beautiful curled spiral decorations resembled the
chambered shells of marine molluscs and were known as snake
stones (Figure 28). Others, called Chedworth Buns, were divided
symmetrically into five sections, looking exactly like petrified sea
urchins.[6] What were the petrified bodies of known and unknown
marine creatures doing buried beneath fields far from the sea?

The standard explanation in the seventeenth century was that
the figured stones were independent creations of God. He had
placed them on the earth for His own mysterious reason, perhaps
as some kind of creation apprentice piece before moving on to flesh
and blood creatures, or simply to remind humanity of His omnipotent power. Yet their distribution remained enigmatic. Why
were they abundant in Oxfordshire or Dorset but hardly ever found
on Dartmoor or in the Welsh hills? Why would God have provided
a lesson written in stone for the men and women of Dorset but
not for those of North Wales?[7]

A few scientists had taken a more radical view. In his *Micrographia*,
published in 1665, Robert Hooke described the microstructure
of, not only living specimens, but also some figured stones and
was astonished to discover that not only did they look like living
specimens to the naked eye, but also under his microscope. He
proposed that the figured stones were exactly what they looked
like, the petrified remains of animals and plants. Although initially

received with a great deal of scepticism, his idea gained momentum, particularly as many specimens showed obvious signs of having been crushed and dismembered in a state that did not appear to be consistent with their deposition by a caring God keen to impress humankind with his omnipotent power.

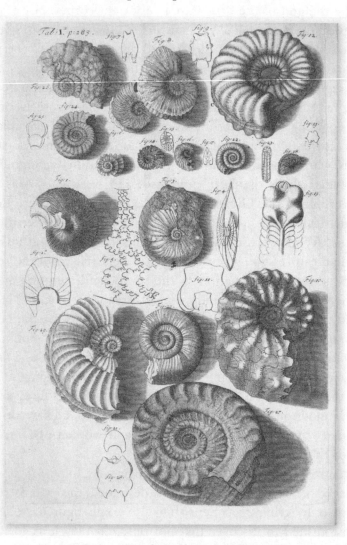

FIGURE 28: Robert Hooke's drawings of figured stones. These were known as snake stones, which we now know to be the fossils of ammonites.

Another problem for the creationists was the discovery of figured stones of creatures unlike any known to science. In 1811, Mary Anning,[8] an amateur collector of figured stones, uncovered the petrified seventeen-foot-long skeleton of a completely unknown sea creature jutting out of the Dorset cliffs. What could God have had in his mind when he made stone ichthyosaurs? Across the Channel, the French zoologist Georges Cuvier (1769–1832) had similarly unearthed what were clearly the fossilised bones of unknown land animals, such as mastodons, mammoths, giant ground sloths and pterodactyls, insisting that they were the remains of extinct creatures. By the early nineteenth century most naturalists were convinced that Robert Hooke had been right. In his hugely influential *Principles of Geology*, published in 1830, the great English geologist and friend of Darwin, Charles Lyell (1797–1875), accepted the principle that fossils were the remains of extinct plants and animals.

Extinction posed a huge challenge to the human-centred creation story. Why would God have created animals for humanity to have 'dominion' over only to later wipe them out? Some creationists argued that flesh and blood versions of stone fossils still flourished in remote and uninhabited regions. However, although a weak case might have been made for, say, a sea creature like the ichthyosaur having escaped detection, it was hard to believe that no one would have spotted a pterodactyl flying across contemporary earthly skies. As the young Alfred Wallace observed on reading a treatise on the classification of animals and fossils according to biblical creationism, 'to what ridiculous theories will men of science be led by attempting to reconcile science with scripture.'[9]

Running alongside the debates about the nature of fossils were also doubts about that other central tenet of creationism, the immutability of species. The French aristocrat and anatomist Georges-Louis Leclerc (1707–88), Comte de Buffon, had noticed vestigial parts in some animals, such as the bones of useless lateral toes in the pig. Why, Buffon asked, would God have fitted animals with useless parts? Buffon thought it more likely that the animals with vestigial limbs had descended from a related species now extinct where the defunct limbs had once been functional.

Buffon was employed as director of the Jardin du Roi in Paris and there employed and mentored the French naturalist Jean-Baptiste Lamarck (1744–1829) who, in 1809, fifty years before Wallace's letter arrived at Down House, published his book *Philosophie Zoologique*, arguing that all species evolve via their inheritance of acquired characteristics. His famous example was an antelope that stretched its neck to reach the highest leaves on a tree and then passed on the acquired characteristic of its stretched neck to its offspring whose further exertions to eat those hard-to-reach leaves would, in turn, be passed to their offspring until, eventually, a giraffe was born.

Most scientists remained sceptical, particularly as acquired characteristics did not appear to be inherited. Blacksmiths were a famous counter-example. Their arms tended to be very asymmetric, with their hammer arm much more muscular than the non-hammering arm. Yet blacksmith's children did not inherit this asymmetry, unless they became blacksmiths themselves. Nevertheless, by the mid nineteenth century, no one had come up with a better theory of the origin of species. When, in 1836, the English astronomer and philosopher John Herschel wrote to Charles Lyell, asking for his thoughts on 'that mystery of mysteries, the replacement of extinct species by others', Lyell replied that God had created each species to be perfectly adapted to its environment but in a process of continuous creation over geological periods of time.[10] Lyell was proposing a kind of gradualist creation.

This origin of species question was therefore a hot topic, even in the relative scientific backwaters of Wallace's Leicestershire. So when, in the 1840s, Alfred and Henry took a break from pinning beetles, they would often discuss the findings and ideas of Buffon, Lamarck, Humboldt, Lyell or Darwin. A determination grew between them that, together, they would seek the solution to the origin of species problem.

## *Amazonian mischief*

In 1845, Alfred's ambitions had to be put on hold after he received news that his elder brother William had died of pneumonia. With five of his siblings now dead, Alfred was forced to assume the role of head of the family and chief breadwinner. He resigned from his teaching job to take up more highly paid work as a surveyor once again.

For the next two years Alfred worked with engineers and railway construction workers, surveying the countryside to find land appropriate for railways. Yet he continued to correspond with Henry, discussing the recent discoveries in natural history or their own ambition of becoming naturalists. By 1847, Alfred had accumulated savings of what he considered to be a small fortune of £100. Coincidentally, in that same year, Charles Darwin inherited £40,000, his share of the family estate. In the autumn of 1847, Alfred wrote to Henry with a plan. They would journey across the globe in the footsteps of Humboldt or Charles Darwin while earning their living as fly-men. Alfred is explicit about his own ambition to 'take some one family [of plants or animals] to study thoroughly, principally with a view to the theory of the origins of species'.[11]

The two met in London to make preparations. Unlike Humboldt, they could not fund their own expedition; neither did they, like Darwin or Halley, have influential connections who could obtain for them a free ride on a Royal Navy ship. Instead, they visited the British Museum and met its butterfly curator, Edward Doubleday, who advised them that the little-explored north of Brazil would be likely to yield rare and valuable specimens. They walked to Kew Gardens and there met the director, Sir Joseph Dalton Hooker, from whom they obtained letters of introduction together with a wish list of rare palms. They even found an agent, the fellow natural-history enthusiast Samuel Stevens, who had recently founded the Natural History Agency. Finally, they pooled their resources to buy berths on the sailing barge *Mischief*; and, following in the footsteps of their hero, Alexander von Humboldt, they set sail for South America.

Alfred never forgot his first vision of the tropics as they sailed into Salinas, Brazil, the pilot-station for vessels bound to Pará, the port of entry to the Amazon basin. He described it as 'a long line of forest, rising apparently out of the water'. They disembarked on the 26 May 1848 into a town whose people were 'every shade of colour . . . from white to yellow, brown and black negroes, Indians, Brazilians and Europeans with every intermediate mixture'. They breakfasted on fried monkey and walked beyond the town into the forest where 'slender woody lianas hung in festoons from the branches, or were suspended in the form of cords and ribbons, while overhead luxuriant creeping plants overran alike tree, trunk, roofs and walls or toppled over palings in copious profusion of foliage.' The pair rushed off to embark on a series of expeditions upriver to the Guaribas rapids and further into the forest where they encountered alligators, vampire bats, wasps and a host of biting insects, forcing them to shroud their heads in netting hung from their wide-brimmed hats. On their first expedition into the jungle, they managed to shoot, net or collect 3,635 specimens of insects, birds and plants, many new to science, which they preserved, packed and shipped back to Stevens in England.

After nine months working together, Alfred and Henry decided that they would be more productive if they separated, with Alfred taking the Amazon tributary of the Rio Negro, even as far as the most southerly point of Humboldt's journeys in Venezuela, while Henry explored the River Solimões. Alfred continued to collect specimens as well as apply his training as a surveyor to map this largely unexplored region of the Amazon river basin. As he travelled upriver by canoe, Alfred met with Amazonian tribesmen who told him forest tales of jaguars, pumas, fierce wild hogs, wild men with tails and the dreadful *curupuri*, the demon of the woods. His encounter with local people instilled in him a deep and life-long fascination and respect for native cultures and customs. He also developed what he described as an 'excited indignation against civilised life'.

FIGURE 29: Frontispiece from Henry Walter Bates, *The Naturalist on the River Amazons* (1863), 'adventure with curl-crested toucans'.

In 1849, Alfred wrote to his family to suggest that his younger brother Herbert join him. Herbert arrived with another young explorer, the botanist Richard Spruce, and the three spent the next two years collecting among the diversity of flora and fauna of the Amazon. Sadly, Herbert died of yellow fever in Pará in 1851 and Alfred suffered a series of bouts of fever or ague, probably malaria.

He recovered but felt debilitated and depressed. Although Henry Bates would remain in the Amazon for another six years, Alfred decided it was time for him to return to England.

In July 1852, Alfred returned to Pará and packed the last of his specimens into boxes that were loaded, along with a live menagerie of birds, monkeys and a wild dog, onto the sailing ship *Helen*, bound for England. Two days out and just after breakfast the captain burst into Alfred's cabin with the news that 'I'm afraid the ship's on fire. Come and see what you think of it.' Alfred could only save his diary and a few pencil drawings of Amazonian fish before he and the rest of the crew were forced to abandon ship and take to small boats. From there Alfred watched aghast as all his precious specimens succumbed to the fire and his terrified animals and plants either perished in the flames or drowned along with the sinking ship. Only a parrot managed to save itself by falling into the sea where it was picked up by a seaman on one of the lifeboats.

After ten days adrift, Alfred's face and hands were blistered by the sun and supplies of food and water in the boats were running perilously low (we are not told what happened to the parrot). Luckily, they were spotted by the crew of a lumbering brig, the *Jordeson*, limping its way to England. The elderly ship could only average two or three knots but after eighty more long days at sea, they eventually disembarked at Deal in Kent, where Alfred joyfully dined with both captains. He was even more cheered when he heard that Stevens had insured his cargo for £200. His reliable agent had also arranged for several of Alfred's letters describing his observations to be published, and had displayed and already sold many of the specimens that Alfred had packed off to England in earlier shipments. So when Alfred arrived in London he discovered, to his surprise and delight, that he had risen from obscurity to being a moderately well-known collector and respected naturalist.

With £200 in his pocket Alfred wasted no time in making plans to set off on a new expedition. This time he chose to head east. In March 1854 Alfred set sail for the Malay archipelago and arrived in Singapore in April, four years before he posted the famous letter to Darwin.

## Constraining life's history: the Sarawak law

Alfred spent his first three months exploring and collecting specimens of the fauna and flora of Singapore before arriving, on 1
November 1854, at the port of Kuching in the Sarawak province
of Borneo where he set up his base. There he hired a fifteen-year-old
Malay boy, called Ali, to cook and help him learn Malay. Ali also
turned out to be an expert shot and skinner of birds and so remained
with Alfred throughout his eight years journeying through the
archipelago.

From his base, the pair paddled a canoe up the Sarawak and
Santubong rivers. Once on shore, he and Ali would spend the day
shooting birds, trapping lizards and netting insects before retiring
to a nearby Dyak village. There he generally slept in a thatched
timber longhouse under rafters decorated with shrunken heads.
Despite their gruesome ornamentation, Wallace enjoyed the
communal lifestyle provided by the Dyak longhouse, which he often
shared with up to two hundred villagers. Best of all, Dyak villages
were surrounded by forest that looked a lot like its Amazonian
counterpart and yet was filled with an astonishing diversity of birds
and insects completely different from Amazonian species. Gradually,
Alfred began to discern a pattern and the germ of a simple idea
took form in his mind. In 1855, he wrote up his thoughts in a
scientific paper entitled 'On the law which has regulated the introduction of new species'. He posted it to his agent who passed it
on to the popular science magazine *Annals and Magazine of Natural
History*, which published it later that year.

In his paper, which I believe should be far more widely recognised
as key to the development of the theory of natural selection, Wallace
begins by insisting that 'during an immense but unknown period,
the surface of the earth has undergone successive changes'. Here
Wallace was essentially restating the conclusions of Lyell's *Principles
of Geology* in order to provide the 'immense' time he needed for
the rest of his story. He next enlisted the evidence of the fossil
record to claim that 'the present condition of the organic world is

clearly derived by a natural process of gradual extinction and creation of species.' Note that, although he uses the term 'creation', he refers to the process 'of extinction and creation' as being 'natural'. Wallace was clearly nailing his colours to a mechanistic creation of species. He went on to marshal nine 'main facts' of natural history that, he proposed, must be accounted for by any theory of the origin of species.

Four of Wallace's facts are geographical. The first two point out that those broad taxonomic groups, such as butterflies or mammals, have a much wider range than narrower groups, such as families or species. For example, butterflies are found right across the globe but the beautiful longwings, or heliconians, family are confined to North and South America and particular species of heliconians tend to be confined to a region of a forest. Wallace's third fact was that closely related species or groups of species tend to inhabit neighbouring territories. Wallace's last geographical fact is that, when similar climates are 'separated by a wide sea or lofty mountains', then the families, genera and species found, say, on one side of the mountain range will be closely aligned with the families, genera and species on the other side. He illustrated this fact – one that Darwin had also noted – with his own observations on the islands of Malacca, Java, Sumatra and Borneo, separated by only a narrow shallow sea.

Wallace then provides another four similar facts but – and most remarkably, as this was entirely novel – refers to distance in time rather than space. From the fossil record, he argues that smaller taxonomic groups, such as ammonites, tend to have a narrower temporal distribution in the fossil record than larger groups, such as molluscs. Also, 'species of one genus, or genera of one family occurring in the same geological time are more closely allied than those separated in time'. For example, more closely related species of ammonites are clustered into adjacent strata in the geological record whereas distantly related species are widely separated. His ninth and final 'fact' was that no species or group occurs more than once in the geological record. In other words, 'no group or species has come into existence twice'.

Even more revolutionary for biology, Wallace then managed to gather all nine of his facts into a single simple proposal or 'law', one of the first of modern biology. In his Sarawak law, as it has come to be known, Wallace proposed that 'Every species has come into existence coincident both in space and time with closely allied species.' This principle is so familiar to us today that it is hard to appreciate its originality in the nineteenth century. We take for granted that, for example, we humans and chimpanzees are 'allied species' that came into existence relatively recently in Africa; whereas we and butterflies are less closely related species that separated from a common ancestor in a much more distant time and place. However, in 1855, Wallace's Sarawak law would have been shocking to most naturalists who believed that chimpanzees, butterflies and all creatures inhabiting the earth had been created in the same week and same place six thousand or so years earlier.

We have already discussed the importance of laws, from Buridan's to Kepler's, Boyle's or Newton's: they are the simplest description of a wide range of phenomena and they act as prediction engines. Wallace's Sarawak law is no exception. First, it is simple and economical. It essentially condenses Wallace's nine facts and multitude of observations about natural history into a single sentence. Like earlier simplifications, such as heliocentricity, observations that were previously arbitrary became consequences of the law. Wallace argues that:

It [his Sarawak law] also claims a superiority over previous hypotheses, on the ground that it not merely explains, but necessitates what exists. Granted the law, and many of the most important facts in Nature could not have been otherwise, but are almost as necessary deductions from it, as are the elliptic orbits of the planets from the law of gravitation.

With Wallace's Sarawak law, Occam's razor had found purchase in biology and the natural world became several notches simpler. Yet hardly anyone noticed.

## A trumpery business

After posting his Sarawak paper to his agent, Wallace planned to follow up with a more substantial project. In a letter to Henry Bates, he revealed: 'That paper is of course merely the announcement of the theory, not its development. I have prepared the plan & written portions of an extensive work embracing the subject in all its bearings & endeavouring to prove what in the paper I have only indicated.'[12]

For reasons that will become clear, Wallace's 'extensive work' was never finished; and, by and large, his Sarawak paper was largely ignored. His agent, Stevens, even advised that several of his customers were of the opinion that he should stick to being a fly-man and leave theorising to professionals. However, Charles Lyell, the most famous geologist in England, did read Wallace's Sarawak paper and could see that it undermined his own favoured theory of continuous creation. His response was to argue that 'there are innumerable reasons connected with the past & future as well as the present which will cause the new species to resemble those which exist or which lately existed.'[13] Lyell's 'innumerable reasons' were of course added complexity. He proposed that when God created a new species he planned for its future and those plans might include, for example, that a single island will later be divided in two.

Yet, whether or not he agreed with the conclusions, Lyell was clearly impressed by Wallace's paper and recommended it to his friend. Darwin appears to have been less interested as he wrote a comment in the margins of his copy that 'it's all Creation with him', completely missing Wallace's assertion that he was aiming for 'a natural process of gradual extinction and creation of species'. Darwin concluded that the work contained 'nothing very new'; yet this dismissive assessment was contradicted by another note written alongside Wallace's claim that fossil species occurring in the same geological time are more similar than those separated in time, writing: 'Can this be true?'[14]

A year later, in April 1856, Charles Lyell and his wife visited the Darwins at Down House. They discussed Wallace's Sarawak paper, prompting Darwin to confide in Lyell that Wallace had earlier written to him directly (the letter has since been lost) asking for his opinion on the Sarawak paper and complaining that he had been disappointed by the silence that had accompanied its reception. Fearing that his friend could be scooped, Lyell urged Darwin to publish his theory quickly. He later noted that he discussed: with Darwin 'Mr. Wallace['s theory of the] introduction of [new] species, [that Wallace claimed are] most allied to those immediately preceding in Time, or that [any] new species was in most geological periods [or strata] akin to those [strata] immediately preceding [it] in Time, [which, if true] seems explained by the natural selection theory'.[15]

Lyell's interest prompted Darwin to write a reply to Wallace's Sarawak letter, assuring him that his paper had indeed been read and admired by the people who mattered, including Lyell. As for his own opinion, Darwin wrote that 'I agree to the truth of almost every word of your paper' and went on to state that:

> This summer will make the 20th year (!) since I opened my first note-book, on the question how & in what way do species & varieties differ from each other. – I am now preparing my work for publication, but I find the subject so very large, that though I have written many chapters, I do not suppose I shall go to press for two years.

Was Darwin issuing a polite 'keep-off my patch' notice? If so, Wallace ignored or more likely missed it. Darwin concluded his letter by requesting that Wallace acquire specimens for him of any domesticated poultry that he might 'stumble upon'.

Meanwhile, back in the Malay archipelago, Alfred sailed south from Borneo to the island of Bali and from there further east to the island of Lombok, crossing the dangerous Lombok Straits, famous for their strong currents and sudden whirlpools. He made the crossing on a schooner manned by a Javanese crew who claimed that 'their sea is always hungry and eats up anything it can catch'. Fortunately, the sea wasn't so hungry that day and, after an exciting ride, Alfred disembarked onto the beach on Lombok and set off to explore. He was astonished to discover that, though it was only twenty miles east of Bali and visible from its shores, the island hosted an entirely different ecosystem with honey-suckers and screeching white cockatoos, bee-eaters and kookaburras, birds that he knew were common in Australia but unknown in the west of the archipelago. Everywhere he looked he saw species identical or related to those he knew to be peculiar to Australia and its neigh-bouring islands. Wallace had inadvertently discovered what is today known as the Wallace Line, the discontinuity that runs right through the Malaysian archipelago between the flora and (particularly) fauna, with the Asian species to the north and west, and Australian species to the east and south. This was surely the most startling confirm-ation of his fourth fact of natural history: that regions 'separated by a wide sea or lofty mountains' develop their own distinct flora and fauna.

From Lombok, Alfred embarked on his most adventurous voyage to date, a sea journey of around 1,500 miles to the Aru Islands, close to Papua New Guinea, in a native prau, a sailing boat a bit like a Chinese junk, accompanied by flying fish and leaping dolphins. On arrival, he was hugely impressed by the native Polynesians with their elaborately carved canoes and headdresses decorated with cassowary feathers. He soon set off with gun and net and managed to bag his best prize so far, a specimen of the magnificent king bird-of-paradise. He noted that this rare but flamboyantly coloured animal is found only deep in the forest far from man – evidence, he claimed, that 'must surely tell us that all living things are not made for man'.

Before leaving, he posted off the first of several scientific papers on the natural history of the Aru Islands and sailed to Makassar in Sulawesi and north into the Moluccas or legendary Spice Islands. In January 1858 he alighted on the island of Ternate, where he rented a house close the beach and under the shadow of a smoking volcano. This would be his home and base for the next three years.

Wallace immediately set off to explore the neighbourhood, hiring a small boat and crew and landing at the bay of Dodinga that separates the northern and southern halves of the nearby Maluku Island. There he rented a small hut and after a short though productive walk he managed to net several unknown insects. However, he soon came down with a fever, probably malaria, and was forced to retire to his cabin where he spent the next several weeks. Confined to his hut, Wallace turned to mulling over the species question that had haunted him since his beetle-collecting days with Bates: how is the fantastic diversity of species generated? His Sarawak law had provided a clue, that related species arise nearby in time and space, but it did not provide a mechanism. In this sense it was closer to Kepler's kinematic laws rather than Newton's causal laws. The missing piece of the puzzle was why and how allied species emerge in the same place and at the same time.

With perhaps his own mortality in mind – it was only seven years since his brother had died of a tropical fever similar to his own – his ideas turned to Thomas Malthus's *An Essay on the Principle of Population* and its grim observation that reproduction always outstrips available resources, leading to an inevitable natural cull on population growth. He now combined this notion with the extensive variation that he, and others, had discovered within any species. Crucially however, he knew, from remarkable experiments that he had actually performed in the forest, that natural variation within a species is heritable. Now, mid-fever, the puzzle unlocked. Allied species emerge in the same place and at the same time because they are derived from the same ancestor by the process that we know today as natural selection, thereby accounting for his Sarawak law. This time, Wallace really had discovered the biological equivalent of Newton's causal laws of motion.

Alfred waited for his fever to subside before returning to his house at Ternate. In three days he jotted down his ideas in the form of his paper entitled 'On the Tendency of Species to form Varieties; and on the Perpetuation of Varieties and Species by Natural Means of Selection'. But who to send it to? His first thoughts were probably to his agent, Stevens, who would surely have dispatched it to an appropriate journal. However, Darwin had told him that Charles Lyell had found his Sarawak paper interesting. This was enough to persuade Wallace to aim higher and send his Ternate paper to Darwin asking that he pass it on to the great man of British science. He posted his letter, the one that arrived in Down House on 18 June 1858, and set off for Papua New Guinea to collect more specimens.

By the time Wallace arrived in New Guinea, Darwin was reeling from the shock of discovering that this hardly known Malay-based fly-man had independently arrived at the same idea that he had been nurturing for at least a decade. Also in June 1858, when Darwin had written to Lyell passing on Wallace's Ternate manuscript, he had promised to write to Wallace to offer to pass his paper on to a scientific journal. In the event, Darwin did not write to Wallace but instead penned another letter to Lyell later that week insisting on his own prior claim to the theory of natural selection and pleading that he

> should be extremely glad now to publish a sketch of my general views in about a dozen pages or so. But I cannot persuade myself that I can do so honourably . . . This is a trumpery affair to trouble you with, but you cannot tell how much obliged I shall be for your advice.

The remainder of this story has been told many times[16] so I do not need to repeat it here. Lyell, together with Hooker and the eminent biologist and anatomist Thomas Huxley, arranged for Wallace's Ternate paper on natural selection to be read at a meeting of the Linnean Society on 1 July 1858, but only after the reading of two 'proofs' of Darwin's priority to the theory of natural selection. The first 'proof' was Darwin's previously unpublished 'sketch' of his theory that he had written about a decade earlier. The second was the copy of a letter Darwin had written to the American botanist Asa Gray in 1857 in which he had outlined some of his ideas.

All three papers were published, in the same order as they had been read, in the proceedings of the Linnean Society in September of that year. Meanwhile Wallace continued working as a fly-man, completely unaware of the intellectual storm that his letter had unleashed. Charles Darwin abandoned his 'big book' to write an 'abstract' of his theory which became his great masterpiece *On the Origin of Species*, published in November the following year. When Wallace eventually learned of the 'delicate arrangement' after the publication of Darwin's *Origin*, he abandoned his plan to write his own 'extensive work' on the origin of species.

Alfred spent another four more years collecting specimens in the archipelago, including the biggest prize of his entire travels, a previously unknown species of bird of paradise, known today as Wallace's standardwing. In fact it was his assistant Ali who had first spotted the animal, exclaiming 'Look here, sir, what a curious bird.' Wallace returned to England in April 1862, with a pair of live birds of paradise in his luggage. The event earned a mention in the *Illustrated London News*. He was immediately elected a fellow of the Zoological Society and received invitations from Charles Darwin, Thomas Huxley and Charles Lyell. Wallace also renewed his friendship with his collecting partner Henry Bates. In the summer of 1862 Wallace enjoyed a visit to the Darwins' Down House in Kent. Both naturalists continued to correspond and remained on excellent terms for the rest of their lives.

Wallace published his own masterpiece, *The Malay Archipelago*, in 1869. As well as describing the natural history of the region, he extolled the virtues of traditional cultures, comparing them to Western civilisation in which, he contended, 'the wealth, knowledge and culture of the few do not constitute civilisation . . . [whose] moral organization remains in a state of barbarism'. He remained a socialist all his life and was a staunch advocate of both land nationalisation and women's rights. Unlike many of his naturalist colleagues, he strongly opposed eugenics.

Darwin died in 1882 and was buried at Westminster Abbey. Wallace was one of the pall-bearers. Aged seventy, Wallace was finally elected a fellow of the Royal Society, forty years older than the age Darwin had been when awarded the same honour. Alfred stated that he would have enjoyed the society's membership more if he had received it while he still possessed the strength to attend its meetings. In 1908, on the fiftieth anniversary celebration of the reading of the Wallace–Darwin papers held at the Linnean Society, Wallace described how natural selection had come to him 'in a sudden flash of inspiration'. Hooker followed Wallace and provided the theory with another puzzle, pointing out that there was no longer any 'documentary evidence' of any of the letters that Darwin had received during the Ternate paper correspondences. Despite Darwin's habit of keeping nearly all his letters, all those written to Darwin by Wallace, Hooker, Huxley or Lyell during that crucial year of 1858, including Wallace's original Ternate paper manuscript, have been lost.

Alfred Russel Wallace continued writing papers on a variety of scientific and social topics. In his book *Man's Place in the Universe*, published in1903, he introduced the concept of astrobiology by reviewing the physical conditions required for organic life in terrestrial ecosystems, concluding that the earth is the only habitable planet in our solar system. In 1906, he published a paper entitled 'Is Mars Habitable?' in which he soundly refuted the astronomer Percival Lowell's claims that Mars is 'inhabited by a race of highly intelligent beings'.[17]

Wallace died peacefully on 7 November 1913. He was buried in Broadstone cemetery in Dorset, where he had spent his last years. His grave was later marked with a fossilised tree trunk found on a Dorset beach. However, perhaps the most endearing tribute to the memory of one of the world's greatest naturalists and co-discoverer of the theory that the philosopher Daniel Dennett has called 'The single best idea that anyone has ever had'[18] was a brief encounter that the American biologist Thomas Barbour enjoyed when collecting specimens in Ternate in 1907. There Barbour chanced upon a 'wizened old Malay man . . . with a faded blue fez on his head'. In perfect English the old man told Barbour: 'I am Ali Wallace.'[19]

Whatever its provenance, natural selection is probably the most Occamist reduction of a multitude of arbitrary facts into a simple law. It depends on the simplest possible mechanism: an unlimited source of heritable variation combined with differential survival and replication. Both Darwin and Wallace had plenty of evidence

FIGURE 30: Alfred Russel Wallace's tombstone in Broadstone, Dorset.

for differential survival and replication but, less than a decade after publication of *On the Origin of Species*, the politician, scientist and writer George John Douglas Campbell (1823–1900), the 8th Duke of Argyll, uncovered a problem with that other ingredient: an unlimited source of heritable variation. In his *Reign of Law* published in 1867, Campbell pointed out that, despite the book's lofty title, 'Mr. Darwin's theory is not a theory of the Origin of Species at all, but only a theory on the causes which lead to the relative success and failure of such new forms as may be born into the world.' Campbell correctly pointed out that Darwin's magnum opus described natural selection acting on pre-existing variation, such as differences in the shape of a finch's beak or the colours on a butter-fly's wing. Yet that process is not creative; it can only select variants that already exist within a population. On its own, it cannot make new variants nor create new species.

The next step in the unravelling of biology's biggest secret was to discover a simple source of novel variation.

# 15

## Of Peas, Primroses, Flies
## and Blind Rodents

There is no more convincing proof of a theory than its power
of absorbing and finding a place for new facts.
                              Alfred Russel Wallace, 1867[1]

Around the same time as George Campbell highlighted the absence
of a way of generating variation in the theory of natural selection,
Fleeming Jenkin, Regius professor of engineering at the University
of Edinburgh, and inventor of the cable car, uncovered another,
and potentially even more serious, problem. In his review of
Darwin's *On the Origin of Species*, Jenkin pointed out that heredity
tended to blend characteristics. Tall mothers and short fathers
generally have children of intermediate size. This drift towards the
mean would, Jenkin insisted, remove the variation that natural
selection depends upon. Not only that, but he argued that 'the
advantage' of any [new] rare variation would be 'utterly outbalanced
by [its] numerical inferiority'. To hammer home his point, and with
the casual racism typical of the latter half of the nineteenth century,
he cited the example of 'A highly-favoured white [who] cannot
blanch a nation of negroes'.[2]

The root of the problems highlighted by Campbell and Jenkin
was that, in the nineteenth century, no one could answer a question
commonly posed by children.

## Why do I look like Daddy?

Between them, Charles Darwin and Alfred Russel Wallace were survived by ten children. I have not been able to find any pictures of the young Wallaces but there are many photographs of Darwin's surviving children and it is easy to see a family resemblance to both Charles and his wife Emma.

Among the many challenges that have confronted our understanding of the world, heredity is surely the biggest. Like begets like, acorns make oak trees and eggs make chickens. Neither acorn nor egg looks anything like an oak tree or a hen, but somehow the secret of making both is contained within each. How are their secrets encoded in the egg or seed? How is the message unwrapped to make an oak tree or a chicken? Most scientists in the nineteenth century fell back on the usual fix of divine intervention allowing heredity to become the last refuge of vitalism. In desperation, Darwin resorted to the discredited theory of acquired (Lamarckian) inheritance that he called pangenesis. In his 1868 book *The Variation of Animals and Plants under Domestication* he proposed that characteristics acquired during an animal's life are transmitted from the body to the gametes (egg and sperm cells) via particles that he called 'gemmules'. The theory was subject to the same criticism (remember the blacksmith's hammer arm) as Lamarck's and failed to convince his critics. Even Wallace eventually came out against it. In the last decades of the nineteenth century, the theory of natural selection nearly suffered the fate of many of the creatures that it described and became extinct.

Yet, two years before Jenkin published his challenge to the theory of natural selection, the solution to the blending problem of heredity had already been uncovered by a little-known Augustinian friar.

## The message in a pea pod

Johann Mendel (1822–84) was born into a peasant farming family in Hynčice, a small Silesian village now in the Czech Republic. At that time, most farmers begat farmers, which would have been Johann's fate if the local schoolmaster hadn't spotted the boy's talents and intervened to persuade the family to use their hard-earned resources to send Johann to the local high school in nearby Troppau (Opava). Johann graduated six years later though not without a struggle as throughout his life he suffered from bouts of what we would today call clinical depression.[3]

Johann next enrolled at the University of Olomouc in Moravia to study philosophy and physics. His sister Theresia paid his fees out of her dowry and Johann earned money for his board and lodgings by tutoring younger students. It was probably at Olomouc where Johann acquired an interest in heredity, as its dean of natural sciences, Johann Nestler, had performed his own animal- and plant-breeding experiments. Theresia's dowry was not however bottomless so to continue his education Mendel entered St Thomas's Abbey in Brno in 1843 as a novice friar. There he adopted the name Gregor. As he later wrote, 'my circumstances decided my vocational choice'.

Gregor Mendel was first trained as a priest and given his own parish but, in a 1849 letter to the local bishop, Abbot Cyril Napp admits of Mendel that: 'He is very diligent in the study of sciences but much less fitted for work as a parish priest.' The abbot sent the scientifically minded friar to the University of Vienna where he studied physics, under Christian Doppler, famous for discovering the Doppler effect. He was also taught botany by Franz Unger, a microscopist who had tinkered with his own pre-Darwinian theory of evolution. In 1853, Mendel returned to Brno.

It is not clear exactly why Mendel decided to study peas. The choice was however consistent with the principles of experimental science established by Galileo, Boyle and others of keeping experimental systems as simple as possible. Peas are easy to grow, have a short generation time and come in several easily recognised and heritable varieties distinguished by, for example, their fruit, that is either round or wrinkled, green or yellow, their height, or the colour of their flowers, which might be white or purple. Like Galileo filing iron balls so they were perfectly spherical and rolled smoothly, Mendel further smoothed his experimental pea model by inbreeding each variety for several generations until the character was stable. To remove another source of variation, he crossed individual pea plants manually so that he knew the exact parents of each cross. As Mendel wrote, 'Experiments with seed characters give the result in the simplest and most certain way.'[4] By this time, Mendel did not need to quote either Aristotle or Occam to justify his preference for simplicity. It had become so much second nature that most scientists were unaware that they were working in that way.

In his paper written around 1865, Mendel describes his intention to shed light on the process of 'artificial fertilization, in ornamental plants, to obtain new colour varieties'.[5] His studies in Olomouc and Vienna had also made him aware of the evolutionary debates rumbling through the corridors of nineteenth-century natural history. He owned, and had clearly read, a German translation of Darwin's *On the Origin of Species*, so in his revolutionary paper Mendel wrote that his experiments to investigate heredity are 'the only right way by which we can finally reach the solution of a question the importance of which cannot be overestimated in connection with the history of the evolution of organic forms'.

To discover how pea characteristics, such as wrinkled peas or purple flowers, are inherited, Mendel crossed plants with different traits; for example, those that produced white flowers with others that produced purple flowers. He expected to see a next generation of peas that were perhaps a pale purple. Instead, the white character vanished as all the pea plants had purple flowers. Mendel next allowed this first generation of peas to self-fertilise and planted the seeds to grow a second generation. When he inspected their flowers, he was astonished to see that the white character had returned, though only in a quarter of the pods. Rather than the expected blending, Mendel obtained approximately whole number ratios of purple to white peas of three to one.

Mendel performed about 15,000 crosses with many different pairs of characters over an eight-year period and meticulously measured and recorded characters in the offspring over several generations. Remarkably, whatever pairs of complementary traits he examined delivered approximately whole number ratios of the trait in the offspring, for example, three times the number of round versus wrinkled peas (3:1) or equal numbers (1:1) or only round peas (1:0). He also noted that for each binary character (round or wrinkled, purple or white) one variant, such as roundness in peas, tended to be dominant in the first generation after a cross whereas the alternative, *recessive* variant (wrinkled) character was hidden until the second generation.

From the perspective of evolutionary theory, the most important result obtained from Mendel's experiments was that characters did not, as nineteenth-century heredity dogma had insisted, blend. Instead, whether dominant or recessive, the pea characters were passed intact through scores of generations. A wrinkly pea from a pod opened in 1863, when Mendel completed his experiments, was just as wrinkly as the first offspring of his crosses in 1855, despite its passage through scores of round pea plants. Heredity might sort but it did not blend. Mendel called the determinants of the unchanging characters that passed down the generations *elementens* but we know them today as genes.

Unlike Kepler, Mendel never told us about the mental anguishes that he probably suffered in attempting to make sense of the data that contradicted so much dogma. His first conclusion was that the whole number regularities that he observed in hereditary patterns must reflect the rather astonishing fact that heredity is discrete rather than continuous. We might today say digital rather than analogue. With his background in the physical sciences, this property must have surprised Mendel as he surely knew that it separated heredity from all other physical parameters, such as speed, mass, momentum, pressure, temperature or acceleration, which vary continuously. Yet genes appeared to know just a few whole numbers: one, two or three.

Mendel read his heredity paper to a meeting of the Brno Natural

History Society on 8 February 1865 and it was published in the following year, only seven years after Darwin's *Origin*, right in the middle of the battle that was raging between the supporters and detractors of natural selection. Mendel's paper could have answered at least some of the critics. A reference to it appeared in *Guide to the Literature of Botany*, by Benjamin Daydon Jackson, which lay on the shelves of the library of the Linnean Society, where the theory of natural selection had first been unveiled. Yet no one involved in the fray appears to have read it.

When Cyril Napp died in 1867, Mendel was elected abbot. Thereafter, he abandoned his greenhouse and devoted himself to administrative duties. He died on 6 January 1884, aged sixty-one, completely unaware that he was soon to become the father of the science of genetics. His greenhouse was dismantled and all his papers burned in the garden of the abbey.

Mendel's experiments answered Fleeming Jenkin's blending criticism of the theory of natural selection. Yet it left the source of the novel natural variation problem, Campbell's objection, unanswered. The origin of new species still remained a mystery.

## Primroses and flies

Hugo de Vries (1848–1935) was born in Haarlem in the Netherlands in 1848. He grew up in an area rich in plant life that inspired him to take up the study of botany at the University of Leiden in 1866. There he read Darwin's *On the Origin of Species* but was not convinced, for all the reasons already mentioned. In 1886, two years after Mendel's death, while walking over a fallow field near Hilversum he noticed a scattering of evening primroses including several anomalous varieties that had not previously been described. He took the seeds back to his laboratory and demonstrated that the anomalous characteristics were not only heritable but also showed whole number ratios of dominant/recessive characters in the offspring. De Vries called the novel variations *mutations* and claimed that they provided the variation needed to establish new species. Searching for similar studies in the literature, he rediscovered Mendel's work. In 1901, he put forward his theory that mutations provide the source of variation that creates new species.

A few years later, in 1907, the American scientist Thomas Hunt Morgan (1866–1945) began a programme to extensively breed the common fruit fly. After breeding thousands of red-eyed flies he noticed a few with white eyes. He demonstrated that the white-eyed character, a mutation, was inherited in a pattern of whole number ratios. He similarly rediscovered Mendel's work and went on to show that mutations give rise to animals beyond the normal range of variation within a species. The subsequent fusion of natural selection with Mendelian genetics became known as the *modern synthesis* or the *neo-Darwinian synthesis*. It remains the cornerstone of the science of genetics, indeed of biology. As the evolutionary biologist Theodosius Dobzhansky insisted: 'Nothing in Biology Makes Sense Except in the Light of Evolution.'[6]

However, in the early decades of the twentieth century, although genes came to be accepted as the units of heredity and the drivers of evolution, nobody knew what they were made of, nor how they worked. The mystery became the last refuge of life's vital principle and even the hand of God. So, for example, in 1911, the French philosopher Henri Bergson (1859–1941) published his *Creative Evolution*, arguing that heredity and evolution were driven by an *élan vital* or *vital impetus* peculiar to living.[7] The subsequent quest to drive the last vestige of scientific theology out of modern science would uncover the secrets of the most extraordinary molecule in the known universe.

### The code breakers

In the next part of our story I will once again be skipping over the surface of many crucial discoveries to highlight only those that are important for our understanding of the role of simplicity in biology. The first step in the exorcism of vital forces from genes was to show that they are made of ordinary chemicals. This had in fact been achieved around the time that Mendel discovered genes, in

1868, by the Swiss chemist Friedrich Miescher (1844–95). While working at Kepler's university in Tübingen, Miescher isolated a biochemical he called 'nucleic acid' from white blood cells and showed that it was composed of hydrogen, oxygen, nitrogen and phosphorus. Miescher never discovered what his new biochemical did, but in 1944 the Canadian-American scientist Oswald Avery (1877–1955) demonstrated that genes are made of deoxyribo*nucleic acid*, what we know today as DNA.

Yet even Avery's identification of the chemical nature of genes didn't help much as no one had any idea how the shape of a pea, the eye of a fruit fly, or the colour of your eyes, all heritable characteristics, could be determined by a chemical with only atoms of carbon, oxygen, nitrogen, hydrogen and phosphorus to work with. Moreover, those characters had to be passed faithfully down the generations, obeying Mendel's rules, but occasionally generating novel variants. This is a very tall order for a chemical that can be purified from living cells and dries into something that looks a lot like paper fibres.

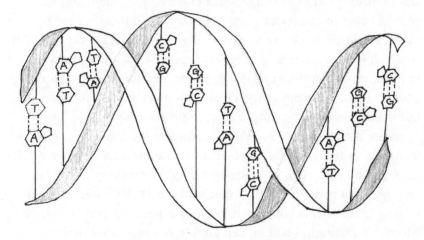

FIGURE 31: The double-helical structure of DNA.

The puzzle was famously solved in 1953 by the Cambridge-based scientists James Watson and Francis Crick, using X-ray crystallography data provided by their King's College, London-based colleague Rosalind Franklin. Their discovery of the double-helical structure of DNA and its ability to encode biological information in genes is probably the most astounding discovery in the whole history of science. The story has been told many times,[8] so I will only highlight the fact that despite the molecule's extraordinary simplicity it was still able to solve the deep puzzle of heredity.

Chief among its simple features is its chemical structure (Figure 31). It is composed of sequences of just four chemical groups called DNA bases labelled A, T, G and C bolted onto a helical backbone like beads on a string. Each strand in the helix is paired with a kind of mirror image of itself, a complementary strand with the simple rule that A pairs with T and G pairs with C. Watson and Crick recognised that these gene letters were codes to make proteins. The DNA to protein coding principle is also simple. Three DNA letters encode each of the twenty protein amino acids that go into proteins. So, for example, GGC codes for the amino acid glycine, whereas CAA codes for glutamine. Proteins make enzymes and enzymes are the molecular factories that make all the other biomolecules inside your cells and those of every animal, plant and microbe that has ever lived on our planet. So the entire biosphere is written in a four-letter code, twenty-two digits fewer than that used to write this book. There is surely no more vivid demonstration of the ability of simple rules to generate extraordinary complexity. Indeed, it has been argued, on quantum mechanical principles, that the genetic code is as simple as it could possibly be.[9]

In the decades following Watson and Crick's discovery, mutations were also shown to be material entities. Chemical DNA bases can be damaged by heat, radiation, strong sunlight or just age. This damage can modify a genetic letter so that, when it is replicated, the wrong coding letter will be incorporated into a gene causing a mutation. Mostly the mutation will be benign, but occasionally it will produce a variant character, such as white rather than yellow flowers in a primrose. If this variation provides an advantage, then natural selection will ensure that its descendants who carry the advantageous new variant will become more numerous. If this happens in an isolated population, then they will found a new species. Yet, if the novel variation makes the creature less fit, then the gene will become less numerous until the mutation is eventually lost from the population. Given the nature of genes and the inevitability of natural selection, evolution becomes as inevitable as an apple falling from a tree.

Once again, as in all science, the genetic mechanism is a model. Like all useful models, it is simple yet has amazing predictive power. The science of molecular biology has harnessed the simple gene model to deliver countless health benefits including new drugs, therapies and novel crops that feed a fast-growing population, as well as the vaccines that as I write are being rolled out to protect the world's population against Covid-19. Yet genes play another, and perhaps paradoxical, role in my argument that life is simple. This time a rather ugly rodent is involved, and some bees.

## The fate of unwanted genes

Eusocial insects (a group that includes bees and ants) are characterised by complex social structures including division of labour, highly complex nests, often a single reproductive queen attended by sterile workers, and sophisticated forms of communication, such as the figure-eight dance of honeybees. Sterile workers at first sight appears to contradict the 'red in tooth and claw' principle of natural selection which would be expected to favour individuals who put their own interests first. Why should a worker ant abandon the option of reproducing itself in favour of helping its sister? The question goes to the heart of a puzzle in biology, especially in relation to our own species: altruism. Contrary to what might be expected from the principle of survival of the fittest, many animals, such as the eusocial insects, share resources and defences, but why?

The English evolutionary biologist William D. Hamilton (1936–2000) proposed a possible solution. Most eusocial insects share a peculiar inheritance system called haplodiploidy, in which males have only one, rather than two copies of all their genes, whereas females have the normal two copies. Applying Mendelian rules to this pattern of heredity ensures sisters share 75 per cent of their genes, rather than the usual 50 per cent typical for peas, humans or other animals and plants. Hamilton did the sums and discovered that a female often has a better chance of passing on her genes by helping her sister, the queen, reproduce, rather than producing her own brood. According to this theory, although worker ants or bees may appear to be altruistic, it is actually their genes that are in charge. The workers and their queen are both slaves to their genes.

What is most remarkable about this theory is that a single tweak to the simple Mendelian pattern of reproduction and heredity can generate vastly different creatures. This is of course a feature of any simple system. Whereas highly complex interconnected structures tend to be robust to perturbations, modifying the rules of a simple system, such as heredity, will reverberate through the entire system to generate large effects. Hamilton published his kin selection theory in 1964 and, although it was initially ignored, it eventually sparked the revolution in evolutionary biology that became known as socio-biology in the 1970s, particularly after the publication of Richard Dawkins's classic *The Selfish Gene*[10] in 1976.

Dick Alexander (1929–2018), curator of the Museum of Zoology at the University of Michigan, was not convinced by Hamilton's theory. As an entomologist and expert on eusocial insects he pointed out that most species with single-gene copy adults, including many beetles, mites, whiteflies and other arthropods, are not eusocial; whereas termites, whose adults, like us, have paired copies of each gene, are. Both ants, termites and bees do however share the habit of building strong defensible communal nests. In a lecture at the University of Arizona in 1976 he presented an alternative theory that eusociality is caused by environmental, rather than genetic, factors. It is a simple theory and, like all simple theories, makes a tight prediction. It predicted that eusociality would arise, even in mammals, wherever there are 'safe or defensible, long-lasting, food-rich nest sites'. He suggested burrowing rodents as possible candidates and the tropics as a likely location as burrows would be sealed tight against intruders beneath sun-baked soil yet have available food in the form of the tubers that store nutrients below the ground to survive bush fires.

After his talk, Alexander was astonished to discover that his theory had already and unwittingly been verified. After stepping down from the podium, a member of the audience, a zoologist called Terry Vaughan, approached him to point out that 'Your hypothetical eusocial mammal is a perfect description of the naked mole-rat of Africa.'[11] Alexander had neither seen, nor even heard of, the naked mole-rat, but Vaughan showed him a dried specimen that he had collected during a sabbatical year in Kenya. Although discovered a century earlier, hardly anyone had studied the animal and, as far as Vaughan was aware, the only person then working on the obscure rodent was a zoologist called Jennifer Jarvis based at the University of Cape Town.

Naked mole-rats are not actually rats. They belong to a family of African rodents known as blesmols that inhabit a similar ecological niche to that of the gophers of North America. Jarvis had been studying the animals since her thesis project in 1967. In the 1970s she had tried to establish a captive colony in her laboratory in Cape Town, but had only ever been able to persuade a single female in the colony to breed. The penny finally dropped when, in 1976, she received a letter from Dick Alexander asking about the animals and telling her about his eusocial mammal theory. Jarvis confirmed that the naked mole-rat was indeed exactly the kind of animal that Alexander had predicted should exist.

FIGURE 32: Naked mole-rat.

Naked mole-rats are widespread in East Africa, where they are sometimes known as sand-puppies. About the size of a small mouse, they are completely hairless with loose skin and tusk-like teeth that are used for burrowing. They remain underground throughout their lives which are spent in dark burrows so, although they do have tiny eyes, they are nearly blind. As well as being of interest to evolutionary biologists, the species has caught the attention of medical researchers as mole-rats never develop cancer and can live extraordinarily long lives of thirty years or more. These physiological features inspired the publication of the naked mole-rat genome in 2011.[12] The data provided clues to genes influencing longevity and cancer; but of particular interest to us is what happens to naked mole-rat genes that fall into disuse.

The researchers discovered that around 250 of mole-rats' genes have picked up so many mutations that they are no longer functional. In a sense, they are dead genes. Despite being non-functional, these *pseudogenes*, as they are called, can still be recognised from signature DNA sequences that identify their former function. Nineteen are involved in vision, for example, one formally encoded for an eye lens protein, another a retinal pigment and a third was involved in transmitting light signals to the brain.

This pattern of genetic mutilation, in the form of repeated mutation, for genes that have fallen into disuse is, perhaps surprisingly, exactly what the neo-Darwinian synthesis of natural selection and mutation theory predicts. As de Vries and others discovered, genes inevitably acquire mutations. However, natural selection predicts that offspring with debilitating mutations will tend to leave fewer descendants than fitter offspring; so defective genes will tend to be lost from a population. For example, a defective vision gene in a mouse will probably end up in the belly of a cat or owl, rather than in any descendants. This process, known as purifying selection, tends to remove mutations that damage genes from the population.

However, whether or not a mutation is damaging depends on the animal's environment. Let's imagine that a landslide buries the entrance to the nest of a family of burrowing rodents. Fortunately, however, their burrow is well provided with an inexhaustible supply of tubers growing down from the top-dwelling plants, so the underground-dwelling animals not only survive but thrive. In their lightless burrow, all animals are blind so they must rely on their other senses such as hearing and smell. In the darkness, natural selection will itself be blind to mutations that damage sight, so purifying selection will no longer work. Without purifying selection, harmful mutations will accumulate until genes involved in sight become non-functional pseudogenes. The sighted rodent will evolve into the naked mole-rat, now so blind that it could never survive above ground.

## Use it or lose it: survival of the simplest

The evolutionary trajectory of the naked mole-rat demonstrates one of the less-appreciated consequences of natural selection: use it or lose it. If a function, such as sight, becomes useless then mutations will inevitably accumulate in its genes until the function is lost. Similar gene decay has been involved in many other evolutionary trajectories. When baleen whales became filter feeders, they gave up biting and so the genes needed to make tooth enamel became pseudogenes.[13] When giant pandas switched from being carnivores to chewing bamboo, they lost the ability to taste umami, the flavour that provides both their ancestors, and us, with a taste for meat. Their dietary switch was followed by the pseudogenisation of the gene encoding their umami taste receptor.[14] Humans have similarly lost a raft of olfactory receptors for odours that we no longer need to smell. And have you ever wondered why your cat doesn't like cake? It's because, after becoming a carnivore, its sweet taste receptor was pseudogenised.[15]

The consequence of this process of mutational accumulation in the absence of purifying selection is that biological functions that are beyond necessity, such as sight in the naked mole-rat, will be removed by a kind of evolutionary Occam's razor. The corollary of the evolutionary razor is that we, and every other living creature alive today is, in terms of function, very nearly as genetically simple as we can possibly be. The 'very nearly' qualification is necessary because evolution may not yet have acted to remove all superfluous complexity, such as the human appendix. Moreover, sometimes it may not be possible for evolution to remove a redundant function, such as male nipples, without extensive remodelling of developmental pathways. Life is indeed simple, but not always quite as simple as it could be.

Use it or lose it is as true for evolution as it is for physical fitness but there may be a more sinister side to this story.

## A deadly simplicity

As I write at the end of 2020, like hundreds of millions of people around the world, I am in lockdown caused by the arrival of a 100-nanometre sphere, around 10 million times smaller than a football, the Covid-19 virus. This tiny particle, completely inert outside a living cell, has brought nearly the entire human world to its knees.

Although it is arguable whether viruses are alive since they are not actually capable of self-replicating, they are the simplest replicating organisms. Rather than replicating themselves, they have opted to dispense with nearly all cellular machinery and instead perform just one task with deadly efficiency: injecting their genomes into our cells where they cajole our cellular machinery to make more viral proteins. These proteins then spontaneously assemble into new virus particles that burst out of our cells to infect other cells to be coughed out of our lungs, shed from our gastrointestinal tract or erupt from skin lesions, indeed any external surface that will lead them to a new host.

In 1977, the biologists Jean and Peter Medawar described a virus as 'a piece of bad news wrapped up in protein'. The bad news is their genome and the protein is the material of the virus shell. Their genome consists of only around 30,000 genetic letters which, when measured in bits, encodes about as much information as a chapter of this book. That information only has one agenda, to replicate. Yet, when expressed in its simplest form, that drive to self-replicate, honed by the simplest law in our universe, natural selection, has evolved the capability to roll over all our plans, hopes, loves, hatreds, thoughts, creativity, ambitions or fears and turn us into mere virus factories. The simple logic of natural selection ensures that, so long as the virus can do this faster than we can kill it, we will be overwhelmed.

No one knows how viruses emerged. They are very different from even the simplest true self-replicators, bacteria, so it has not been possible to trace their evolutionary lineage. One theory is that they just arose by the chance coming together of proteins and nucleic acids within cells. A more likely scenario, in my opinion, is that they are the endpoint of the path of the biological Occam's razor, first making pseudogenes out of genes and then eliminating all superfluous genetic information except the minimal rump needed for self-replication, a virus. Whatever their origin, they are the most definitive demonstration that life is, sometimes, diabolically simple.

# PART IV

## The Cosmic Razor

# 16

## The Best of All Possible Worlds?

HEISENBERG: Nature leads us to mathematical forms of great simplicity and beauty . . . You must have felt this too: the almost frightening simplicity and wholeness of the relationships which nature suddenly spreads out before us . . .

EINSTEIN: . . . That is why I am so interested in your remarks about simplicity. Still, I should never claim that I truly understood what is meant by the simplicity of natural law.

Werner Heisenberg in conversation
with Albert Einstein, 1926[1]

When we left physics to explore biology in Chapter 13, nineteenth-century scientists had already made enormous advances in their application of simple laws to both terrestrial and heavenly motion. Indeed, physicists claimed that physics was, pretty much, complete. Yet at a meeting of the British Association for the Advancement of Science towards the close of the nineteenth century, the Northern-Irish physicist Lord Kelvin (1824–1907) warned that this rosy assessment was contradicted by 'two small clouds' on the horizon, two problems that physics had yet to solve. Remarkably, the resolution of these two clouds provoked two revolutions that overturned most of the certainties of nineteenth-century physics.

Both of Kelvin's clouds concerned the nature of light. Nearly a century earlier, Thomas Young (1773–1829) had demonstrated that light behaves like a wave, similar in principle to water or sound waves. Waves need a medium such as water or air to move through.

Young had demonstrated that light travels through a vacuum, as it must do to reach us from either the sun or a distant star. What could be moving in a vacuum to transmit light waves?

No one had any idea. In desperation, scientists resurrected Aristotle's heavenly aether. You will remember that, in order to make his 'whatever is moved is moved by another' theory of motion work, he filled the spaces between objects with a plenum that, in the heavens, was composed of aether. Although largely abandoned after Boyle's demonstration that nature does not abhor a vacuum, Young's experiments revived Aristotle's aether plenum idea as an entity to fill the void to provide a medium for light waves. Moreover, the aether had the additional advantage of filling that explanatory gap in Newtonian physics by providing a frame in which an object's speed or acceleration could always be measured. Newton had been content to allow God's eye to provide that frame, but, by the late nineteenth century, physicists were more comfortable with a secular aether.

Lord Kelvin's first puzzle was to do with the nature of that aether. If it is some kind of invisible substrate for light waves that fills all of space then it should be possible to measure the speed of an object relative to the aether, just as you can measure the speed of a boat relative to the water in the sea. The task is of course much more challenging, as light waves move at around 300 million metres per second, massively faster than any object on earth. However, in 1887, the American scientists Albert Michelson and Edward Morley came up with the clever idea of measuring light speed relative to the fastest object on earth which, relative to the sun, is the earth itself. The earth spins on its axis at around 447 metres per second as well as orbiting the sun at around 30,000 metres per second. The team realised that, rather like the Doppler effect, the speed of light should be different if measured in the direction of the earth's motion, i.e. through the aether, or perpendicular to that direction.

Yet it didn't. Despite their best efforts, Michelson and Morley always detected exactly the same speed for light, irrespective of whether the earth was moving towards or away from the direction of the light waves. That did not make sense with an aether wave theory of light, something Lord Kelvin found disquieting.

The second of Kelvin's 'clouds' was a problem with classical thermodynamic theory. Earlier in the nineteenth century, Maxwell and Boltzmann had combined Carnot's theory of heat transfer with Newtonian mechanics to derive the modern theory of thermodynamics or statistical mechanics. This envisages matter made up of trillions of atoms whose random motion gives rise to heat. When light interacts with matter it can be absorbed to increase the thermal energy – speed – of atoms. Alternatively, the motion of atoms may be slowed by emission of light energy. Classical thermodynamics worked pretty well to account for the data but failed to account for the spectrum of radiation emitted from an object known as a black body that absorbs all the light that falls on it. A hole in a wall into the darkest of rooms is a good approximation of a black body but it is easier to obtain a sense of this problem by shifting to the more familiar realm of sound.

Imagine that you have a grand piano. Further imagine that you pick up a sledgehammer and swing it into the piano to give it an almighty whack (no actual grand pianos will be harmed in this thought experiment). The piano is full of strings and the impact will set all of them vibrating at all possible frequencies to generate a cacophony that slowly dies down to a faint hum. When black bodies are heated – the molecular equivalent of absorbing a big whack from a sledgehammer – instead of emitting light at all possible frequencies, they emit light in a narrow band that depends only on the black body's temperature. It is as if, after whacking your grand piano, it only emitted a middle C note, which shifted to a D or E if you performed the whacking experiment in a warmer room. Kelvin noted that this, like the constancy of the speed of light, is very odd.

## Relative simplicity

The story of the resolution of Kelvin's 'small clouds' has been told in many wonderful books,[2] so I will focus only on highlighting the role of simplicity. Both involve the work of an extraordinary Bern-based patent clerk.

Albert Einstein was born in 1879 in Ulm, Germany, the son of Hermann Einstein, a salesman and electrical engineer, and Pauline Koch. After a difficult education, he studied physics and mathematics at Zurich but was unable to secure even the most basic university teaching job after he graduated. Luckily, in 1902, a friend of his father offered Albert a position as technical expert in the Swiss Patent Office in Bern. Although the work was routine, Albert was very happy with his job. He later claimed that it was in 'that worldly cloister where I hatched my most beautiful ideas'. One of his most beautiful ideas was a possible resolution of Kelvin's speed of light through the aether cloud. Yet Einstein's interest in the problem did not have its origin in the Michelson and Morley experiments. In fact, he appears to have been unaware of their work. Instead, Einstein was troubled by another odd fact that Kelvin had failed to notice concerning the behaviour of electrical machines.

The origin of the problem lay in the work of the Scottish physicist James Clerk Maxwell who in 1865 discovered a set of simple equations that described the behaviour of both electricity and magnetism in terms of *fields*. In physics, the term is used to describe volumes of space that cause objects to move. So the motion of an apple falling from a tree is caused by the earth's magnetic field; the attraction of a compass needle to the North Pole is caused by the earth's magnetic field; and the motion of a thread of cotton towards amber is caused by its electric field. Maxwell found a single set of equations to describe both electricity and magnetism in terms of fields, demonstrating that both are aspects of a single *electromagnetic* field. In fact, what appears to be an electrical force to a moving observer is experienced as a magnetic force by a stationary observer and vice versa. In writing his equations, Maxwell thereby unified the force of the lodestone with the force of amber as *electromagnetism* to achieve the first great unification, and thereby simplification, of classical physics.

Maxwell's equations achieved probably the most important unification in the whole of physics and thereby made the world hugely simpler. Yet that unification had taken place several decades before Einstein's ponderings in the patent office. What intrigued Einstein was an even more startling simplification lurking in Maxwell's equations. They predicted that the field around an electrically charged object oscillating in space will generate an oscillation in the surrounding electromagnetic field that will radiate away from the object at a speed of around 300 million metres per second. Maxwell recognised that speed. It was the speed of light in a vacuum. His stunning conclusion was that light is a ripple in the electro-magnetic field generated by oscillating charges, which we now know to be electrons, within matter.* Maxwell's unification swallowed not only electricity and magnetism but also the light that illuminates the universe. Light was an aspect of the single electromagnetic force.

Between inspecting patent applications in the Bern office Einstein considered the relationship between light and electricity. Both were highly topical. The Industrial Revolution had, by this time, switched from steam to electrical power and Einstein's father Hermann and his brother Jakob had founded Einstein & Cie., an electrical engin-eering company operating in Munich. Einstein's office was, by this time, lit by electric light bulbs, commercialised only two decades earlier by Edison and Swan. Many of the inventions submitted to the Bern office were electrical machines. The question that Einstein mulled over while sorting through the patent applications was, if he were to design an electrical machine based on Maxwell's equa-tions what value should he use for the speed of light?

Remember Galileo's ship where birds, fish or sailors in the cabin might be completely unaware that they were moving. If they had carried an electrical machine on board, then when calculating how it works would they use the speed of light relative to the cabin walls, or the speed of light relative to the shore that was receding away from the ship? If light speed, like every other speed, is rela-tive then the laws of physics would be different for different observers depending on how they move. Einstein found this idea unsettling.

---

* We now know that the oscillating charges that generate light are electrons hopping between atomic orbitals.

## *Rebuilding physics from the bottom up*

William of Occam stripped medieval scholastic philosophy, and its theological science, down to the simple premise that God is omnipotent, and then examined the consequences. Centuries later, René Descartes dismantled and then rebuilt Western philosophy from his simple conviction that we can only know with certainty the premise that 'I think therefore I am'. Einstein took a similar approach to the science of physics. He was convinced of the universal validity of Maxwell's laws, and reasoned that, to make them universal, light speed had to be the same for all uniformly moving observers.* Unlike every other object in the universe, light, he insisted, cannot obey Galilean relativity.

This seems like a simple statement, but its implications are startling. To get a sense of how odd it is, we will perform what Einstein called a thought experiment, or *gedanken*, by imagining that the waves of the sea are behaving like light waves. We will further imagine that you are about to board a speedboat moored to a jetty at the western end of the Venetian port of San Marco, not far from Galileo's university town of Padua. There, thanks to a freak current in the Adriatic, the waves of the sea happen to be flowing parallel to the coast from east to west along the bow of your boat.

While you and a friend, called Alice, are standing on the jetty off San Marco, you observe the waves rolling alongside your boat taking 2 seconds to travel from stern (rear) to bow. You know that your boat is 10 metres long so you can easily calculate the speed of the waves to be 5 metres per second. This will be your watery equivalent of light speed that must be the same for all observers. You board your boat and, like Galileo, take on board a birdcage with several fluttering budgerigars, a fishbowl where a handful of goldfish are swimming, but also a small dog, Pad, named after Galileo's favourite university.

---

* Einstein's first theory of relativity excludes accelerating objects, that's what made it 'special'.

Pad enjoys a game of running along the deck, keeping pace with the waves. He is a very nifty little dog and can just about keep pace with the fast-flowing waves so that it takes him 2 seconds to run from stern to bow. Alice, watching from the shore, also measures 2 seconds for Pad's sprint. You wave goodbye to Alice, start the engine and setting the boat's speed to a steady 2 metres per second, motor out of the harbour, in the same direction as the waves. You hear the engine roar and see the spray jetting away from the stern. When you have reached a steady speed, you check on your menagerie and confirm that, just as the great Italian scientist predicted, the birds are flying and the fish are swimming unperturbed by the motion of the boat relative to the shore. As Galileo insisted, their motion relative to the boat is all that matters. Speed is relative.

Travelling at 2 metres per second in the same direction as the waves you would expect that the relative speed of the waves along your bow would now be just 3 metres per second (5 minus 2). Pad should be considerably less taxed in his game of keeping up with those waves. Yet, when you let him dash to the stern, he has, once again, to set off at full pelt to reach the bow along with the waves. Has Pad slowed down? To give him a rest, you push the throttle forward until you are travelling at 4 metres per second. At this speed, when you have nearly caught up with the waves, you expect the relative speed of the waves against your bows to be just 1 metre per second. Pad's walking pace should be sufficient to keep pace. Yet he is still straining to keep up with those speedy waves that continue to travel from bow to stern at a speed of 5 metres per second. Perplexed, you look towards the shore and can see San Marco's bell tower disappearing into the distance just as expected. Yet, from your perspective against the waves, you are entirely stationary, just as in the harbour, and San Marco is speeding away from you at a speed of 4 metres per second. Perplexed, you push the throttle forward to your full speed of 5 metres per second, the same as the waves. You should now be keeping pace with these regular waves. Yet, Pad still has to race at his top speed to keep up with them as they stubbornly roll past your bows at 5 metres per second. Once out to sea, far away from any visible land, you can

push the throttle as little or far as it can go but it makes no difference to your progress through the waves. According to the roar of
the engine and the spray from the rear of the boat, you are travelling at top speed, yet, from the perspective of your motion against
the waves, you are completely stationary, marooned in a monotonous
rolling sea that continues to roll past your bow at 5 metres per
second. The motion of these light-like waves stubbornly refuses to
obey Galilean relativity.

The situation is even odder from the shore. Alice has brought
along a replica of Galileo's most powerful telescope so she can keep
an eye on your progress from the jetty. She sees Pad racing along
the deck, but she notices something curious. As your boat acceler-
ates from 0 to 2 and then 4 metres per second, Pad's progress along
the deck appears to slow. Instead of taking only 2 seconds to race
along the deck, he now takes nearly 4 seconds and seems to be
running in slow motion. When your boat has reached full speed
of 5 metres per second, both you and Pad appear to be stationary,
frozen in a moment of time. What is going on?

The answer is that the water waves in our thought experiment
are now behaving like light waves that have the same speed to all
observers. Since Pad's trotting along the deck is keeping pace with
those waves, both you on deck, and Alice standing on the jetty, can
use the dog's motion to measure light speed and you must both
arrive at the same answer in order for Maxwell equations to be the
same for both of you. On shore, this was not problematic, but once
there is relative motion between you, everything changes.

Consider when you are travelling away from Alice at 4 metres
per second. From your perspective (a, in Figure 33), nothing has
changed from the situation when the boat was moored to the jetty.
Pad races the 10 metres along the deck travelling at the light-like
speed of 5 metres per second taking two seconds to reach the bow.
However, from Alice's perspective (b, in Figure 33), the distance
Pad has travelled now has two components. First, Pad races 10
metres along the deck. However, in that time, the boat has also
travelled 8 metres further away from the jetty. From Alice's perspective Pad has travelled a total of 10 plus 8, 18 metres. If Alice were

experiencing the same time as you, then Pad would have run those 18 metres in just 2 seconds at 9 metres per second and thereby faster than our surrogate for the speed of light. To keep faith with Einstein's insistence that light speed is the same for all observers, something has to give. It can only be time.

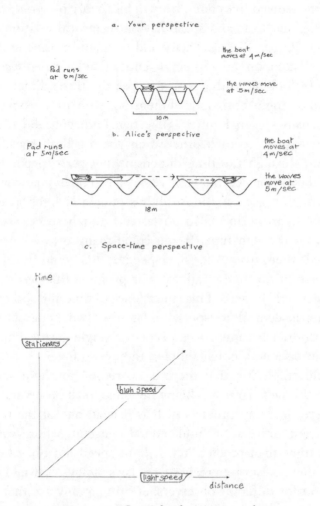

FIGURE 33: Special relativity on a boat.

Since Pad is keeping pace with light speed then, from Alice's perspective, he must take 3.6 (18 divided by 5) seconds to travel the entire 18 metres he has to cover the distance from stern to bow. The event that, for you, took just 2 seconds, must take 3.6 seconds in Alice's experience. You are now experiencing the same event, but from different time frames. When you speed away at full light speed throttle then, from Alice's frame, your time stands still.

Einstein resolved this conundrum in his theory of *special relativity* by insisting that time and space have a reciprocal relationship to each other. Rather like electricity and magnetism, time and space became two components of a single entity that he called *space-time*. One way of looking at this is to collapse the three dimensions of space into a single horizontal dimension, with time as a vertical time dimension (c, in Figure 33). When both you and Alice are stationary, relative to each other, then you are both travelling at light speed through the time dimension but zero speed through the space dimension. As your relative speed increases, your combined speed through space-time must always remain at light speed (the length of the arrow in c). To go faster through space, you must travel slower through time. If you could manage to accelerate to light speed* then, from Alice's perspective, you could travel round the universe in no time at all, as your progress through the time dimension would be zero. This equivalence of time and space within space-time is central to special relativity. Two apparently very different entities became two aspects of a single entity. The world once again became a notch simpler, though stranger.

I should emphasise that there is no reason for light speed to behave in this peculiar way – being the same to all observers. There is no deeper law that predicts it. It is instead one of the building blocks of our universe, a fundamental constant, whose value we observe rather than predict. Yet, if light speed did obey Galilean relativity then, as Einstein realised, the laws of physics would indeed be different for different observers, creating a universe that would be much more complex than our own. This very odd constancy of light speed is how our universe keeps it simple.

---

* This is actually impossible as special relativity prevents any object with mass accelerating to light speed.

Einstein's special relativity paper was one of four that he published in his annus mirabilis of 1905. They established his reputation as one of the greatest physicists of his age and earned him several job offers, prompting him to leave the Bern patent office for successive appointments at the University of Bern, Zurich, Prague and at the Humboldt University of Berlin. Nevertheless, throughout the next twenty years or so, two limitations of special relativity nagged at him: the theory failed when it came to accelerating objects or objects subject to gravity. Also, like us (Chapter 11), he found it curious how, when using Newton's laws to calculate the gravitational acceleration experienced by a falling object, you must first multiply by an object's mass and then divide by that same mass.

## Einstein's Razor

Nature is the realisation of the simplest conceivable mathematical ideas.

Albert Einstein, 1933[3]

For a decade, Einstein struggled to incorporate acceleration and gravity into his relativity theory. Another German physicist, Max Abraham, was also searching but Einstein was scornful of Abraham's approach, which was to look for simple or elegant solutions, writing that 'I was totally "bluffed" by the beauty and simplicity of his [Abraham's] equations'. Einstein went on to blame his failure on 'what happens when one operates formally [looking for elegant mathematical solutions], without thinking physically'.

Einstein's own approach was to construct equations that were compatible with as many observations as he could, however complex they became. Only later did he even check that the equations he had constructed were mathematically sound. He continued ploughing through one complex equation after another and each time, when he got to the stage of checking the mathematical validity of his theory, one after another, they failed.

At this stage in his career, Einstein was eschewing Occam's razor in favour of what is called *completeness*: the incorporation of the maximum amount of available information into a model. You may remember that this is the same issue that was at the root of the debate between myself and Hans Westerhoff about the role of Occam's razor in biology. Hans, like Einstein at this stage in his career, argued for completeness. Yet, as models become more complex, the number of alternatives grows exponentially: consider how many shapes you can make with six, sixty or six hundred Lego bricks. After many years of fruitlessly searching through the vast space of possible models, Einstein eventually changed tack by adopting the approach, for which he had soundly condemned Abraham, of first 'operate[ing] formally, without thinking physically'. He embraced Occam's razor to accept only the simplest and most elegant equations and only later tested them against physical facts. This time he struck gold. In 1915, he came up with 'a theory of incomparable beauty', his general theory of relativity.

The insight that led to the genesis of the theory of general relativity was Einstein's realisation that gravity and acceleration are indistinguishable. Their equivalence, now known as the *equivalence principle*, is something that we become aware of whenever, for example, we take off in an aircraft and feel our weight shifting from our bottom to our back. Einstein realised that they feel the same because they are the same and should thereby be described by a single set of equations. This insight brought him to his theory of general relativity and its by now familiar description of the warping of space-time by massive objects, such as stars and planets. Gravity became an apparent acceleration that is experienced as travelling in an ellipse or parabola in space but as a straight line in space-time. As something distinct from the contours of space-time, gravity became an entity beyond necessity and the universe became another notch simpler.

Einstein also noted that his new conception of gravity solved that puzzle of why mass enters Newton's equation only to be cancelled out when it comes to calculating the acceleration due to gravity. In Einstein's theory of general relativity, gravity is an acceleration provided by the warping of space-time, rather than a force.

A falling object's mass therefore does not enter the calculation for the rate of its fall. In the general theory, gravity is therefore known as a *fictional force*, not a Newtonian force.

The success of the general theory prompted Einstein to reverse his opinion about simplicity and elegance and he thereafter always insisted that the search for mathematical simplicity is fundamental. He advised that: 'A theory can be tested by experience, but there is no way from experience to the construction of a theory.' Here Einstein is essentially stating the inverse problem, that it is easy to go from a simple system (equation) to calculate its complex outputs but usually impossible to do the inverse. He goes on to argue that 'equations of such complexity . . . can be found only through the discovery of a logically simple mathematical condition that determines the equations completely or almost completely.'[4] Chastened by his experience, he thereafter always put his trust in simplicity.

## *Was Ptolemy right after all?*

Before leaving relativity theory, we will revisit our Alexandrian friend Ptolemy to take another look at his remarkable geocentric system of epicycles, eccentrics and equants. As we have already noted, despite its Byzantine complexities, it worked surprisingly well, which brings us back to the question that we have encountered several times in models by Ptolemy, Copernicus or in the phlogiston theory: how can a wrong model get so much right?

The answer is that Ptolemy wasn't wrong; he was just over-complicating. In general relativity, it is perfectly legitimate to take any point in the universe as the centre of your system and perform your calculations accordingly. However, some viewpoints are more complicated than others. As the most massive object by far in our solar system, the sun's influence outweighs all the other gravitational influences. Placing it at the centre of the system is the equivalent of hopping from the shore to the deck of Galileo's imaginary ship to simplify the observed motions of all the objects on board. It is perfectly legitimate to hop metaphorically back to the shore to place the earth as the centre of your system, but you need to draw a lot more circles (see Figure 34). Ptolemy wasn't wrong, and it is a tribute to his genius that he managed to create a system that worked, but he over-complicated matters and the application of Occam's razor led to simpler solutions.

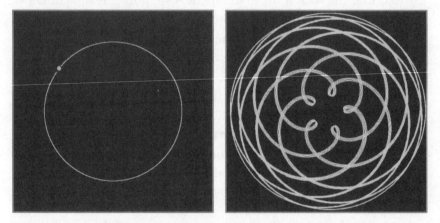

FIGURE 34: Heliocentric (left) versus geocentric (right) perspectives of the orbit of Venus over a thirty-two-year period. The sun is at the centre of the heliocentric orbit and its orbit is the perfect circle in the picture on the left. The Earth is the centre of the geocentric orbit.

Much of physics, indeed of science, involves finding the right perspective to make the world simpler. Special relativity achieved that by unifying space and time to eliminate the first of Lord Kelvin's clouds. Another change of perspective blew away the second of his clouds to reveal a simpler, yet even stranger universe.

# 17

# A Quantum of Simplicity

... man is a quantum ...
William of Occam, c.1320[1]

In 1874, a twenty-year-old student named Max Planck (1858–1947) visited the University of Munich to discuss the possibility of a career in physics. The professor, Philipp von Jolly (1809–84), advised him against it, saying that there was nothing new to discover in the field and he should choose another discipline. Undeterred, Planck enrolled at the university and, in 1877, moved to Berlin to study for a PhD at its Friedrich Wilhelm University where he became interested in thermodynamics. After being appointed professor of theoretical physics in thermodynamics at the University of Berlin in 1900 Planck decided to tackle one of Kelvin's clouds – the failure of thermodynamic theory to account for the spectrum of light emitted from the atoms inside black bodies. Planck discovered an equation that correctly predicted the observed spectrum, yet its implications were startling. Thermodynamics is based on the principle that atoms move randomly at a range of speeds so, when they slow down, they should emit light at a continuous range of frequencies. Yet Planck's equation implied that the light energy from black bodies was instead released in minute packets of energy at discrete frequencies. Planck termed these packets of light energy *quanta*, from the Latin word for a portion or quantity.

Once again, there have been many great books written about quantum mechanics and it is impossible to do this bedrock of twentieth-century physics justice in only a handful of pages. So I will once again focus only on those aspects of quantum mechanics that illustrate the role of simplicity, and a good place to start is to approach this strange science from the perspective of Occam's razor's bedfellow, nominalism. You will remember that the central principle of Occam's nominalism is that abstract notions, such as fatherhood, exist only as words or ideas in our mind, or *ficta*, not as real things in the world. For this reason, Occam insisted that they should be eliminated from philosophy and science.

But what is real and what is abstract in science? We have already discussed how concepts, such as motion, are relative such that an object moving in one frame may be stationary in another. On these grounds, Occam insisted that motion, or impetus, cannot be a 'thing'. Similarly, even gravity is only a fictional force in general relativity. So what is real?

Imagine that you are standing at the edge of an ice rink and wish to precisely measure the location of your friend and ice skater, Alice, who is standing still on the ice. To make the task harder, the lights have all been switched off so that you cannot see her. Fortunately, you have brought along a bag of luminous bouncy balls. To locate Alice, you randomly fling the balls into the darkness. Most fly unimpeded across the rink but you are able to catch a few that bounce back, presumably after hitting Alice. You note the position and direction of your throw and your catch and, applying the principle of triangulation, you can locate Alice's exact position in the dark.

However, as Newton insisted, every action has an equal and opposite reaction. When the ball strikes Alice, a quantity (quantum) of momentum will be imparted to her body causing her to be pushed backwards. Her position after the measurement will not be the same as it was before the measurement. In the macroscopic world, the answer to this conundrum is obvious, you simply turn on the lights to locate Alice precisely.

However, the German physicist Werner Heisenberg (1901–76) realised that, if Alice was a fundamental particle, such as an electron, then even the softest touch by a particle of light, or photon, will always deliver some small quantity of momentum and thereby shift its position.* This realisation led Heisenberg to formulate his famous uncertainty principle which insists that the uncertainty in the momentum of a particle, multiplied by the uncertainty in its position, will always be greater or equal to half the value of another fundamental constant (like the speed of light) of nature, known as Planck's constant. This constant has a very tiny value† and can be ignored for macroscopic objects, such as skaters, but it places a fundamental limit on the precision with which we can even know the microscopic world.

Of course, you could imagine that, despite your lack of knowledge, an electron's precise position is just as real as Alice's position in the dark. However, unlike the macroscopic world, there is no conceivable equivalent of 'turning on the lights' in the quantum world, as light is made of photons and they will impact whatever we want to measure. This provokes a similar question to the one William of Occam asked about the reality of Plato's Forms, Aristotle's universals or motion itself: how real can precise position or momentum be if it can never be measured?

Real things must have an impact on the world. That must surely be our minimum criterion of real. Unreal stuff, like Forms, universals, ghosts or demons, do not have an impact. This is how we know that they are mental, rather than physical objects, or *ficta* as Occam would have described them. If precise particle position does not influence the world (if it did, then it would be measurable) then William of Occam would insist it was no more real than a Platonic triangle, the essence of 'fatherhood' or Henry More's 'knowing spirit'. From the perspective of his nominalism, 'precise position' is just the name that you give to an abstraction that exists in our mind and in our models, a fiction that corresponds to nothing out there in the world. It is thereby an entity beyond necessity and should be eliminated from science.

Quantum mechanics does exactly this. Differences between

* The principle applies to the combined uncertainty of pairs of complementary measurements, such as position and momentum, or energy and time.
† $6.626 \times 10^{-34}$ m² kg per second.

different energy states that are too small to measure are just not considered real. So energy can only be emitted in tiny packets that are measurably different from one another: quanta. Similarly, whereas thermodynamics allows particles to vibrate at a continuous range of frequencies, quantum mechanics insists that only frequencies that are measurable are allowed. It is this *quantisation* that leads to the peculiarities of the black-body radiation and Planck's equation.

However, the strangeness of quantum mechanics is not limited to energy quantisation. Quantum-level uncertainties materialise as counter-intuitive properties of particles that exist in multiple places at once, travel through classically impenetrable barriers or spin in two different directions at the same time, simply because it is not possible to prove that they don't. Similarly, particles may possess spooky connections that reach across time and space, just because Heisenberg's uncertainty principle tells us that we cannot prove that they do not possess such connections.

The proof of the pudding is in, as always, the utility of the science. Quantum mechanics has made some of the most accurate predictions in the entire history of science, as well as delivering novel technologies from lasers to computer chips, GPS, MRI scanners or mobile phones. Probably not far in the future are even more transformative technologies, such as superfast quantum computers or quantum teleportation. Perhaps even most surprisingly, life also appears to be proficient at exploiting this strange and counter-intuitive realm, as outlined in my own book,[2] and a later book I wrote with Jim Al-Khalili.[3]

One of the greatest successes of quantum mechanics is its uncovering of the strange realm of subatomic particles. Yet, on its first encounter, it uncovered, not simplicity, but a jungle.

## *Opening up the atom with quantum mechanics*

Although the idea of atoms goes back at least as far as the ancient Greeks, in the opening years of the twentieth century scientists were still debating whether they were a useful abstraction or real. One of Einstein's four papers of 1905 pretty much settled the matter by demonstrating that the erratic (Brownian) motion of particles, such as microscopic pollen grains, suspended in water, only makes sense if the pollen grains are being knocked about by collision with invisible atoms of water.

When Democritus came up with the idea of atoms, he conceived them as tiny indivisible particles of matter. That notion survived the swapping back and forth between the plenum theory and atomism in the ancient, medieval and modern world up until the late nineteenth and early twentieth century. Then, around the time that Einstein was writing his paper on the reality of the atom, Henri Becquerel (1852–1908) and, independently, Marie (1867–1934) and Pierre Curie (1859–1906) discovered radioactive decay of atoms into smaller parts. Further experiments by Ernest Rutherford (1871–1937) demonstrated that atoms consisted of a tiny positively charged nucleus, surrounded, at a distance 100,000 times the diameter of the nucleus, by a cloud of negatively charged electrons. Later experiments established that atomic nuclei are composed of both positively charged protons and electrically neutral neutrons, delivering our familiar and relatively simple picture of the atom that you might see printed on a T-shirt.

This simple picture did not last. Several physicists noticed that radioactive beta decay did not add up: something was missing. In a letter addressed 'Dear radioactive ladies and gentlemen' written to his colleagues at the Swiss Federal Institute of Technology in 1930, the quantum physicist Wolfgang Pauli (1900–58) predicted the existence of yet another fundamental particle, with zero charge, like the neutron, but possessing much less mass. His Italian colleague Enrico Fermi (1901–54) dubbed the new particles *neutrinos*, meaning little neutrons, bringing the fundamental particle count up to four. In 1928 the English physicist Paul Dirac (1902–84) managed to fuse quantum mechanics with special relativity theory, but the equations demanded that each particle be accompanied by an *antiparticle*, a kind of mirror-image particle with an opposite charge. It was not long before the electron's positively charged sister the *positron* was discovered as a vapour track in a cloud chamber.*

Cloud chambers became the favourite tool of particle physicists, who took to hauling them up mountains to catch cosmic rays that would otherwise be absorbed by the earth's atmosphere. Their experiments revealed showers of new particles fired out of deep space. In 1936, the *muon* was discovered with the same charge as the electron but about a hundred times its mass. Its discovery prompted the American physicist Isidor Isaac Rabi (1898–1988) to quip, 'Who ordered that?' The fundamental particle count soon rocketed into double figures as new particles continued to leave trails in cloud chambers. The situation only worsened when particle accelerators came online in the 1950s and a host of new particles sporting exotic names, such as *pions*, *kaons* and *baryons*, burst out of high-energy particle collisions. Witnessing the zoo of supposedly fundamental particles, Pauli exclaimed: 'Had I foreseen that, I would have gone into botany.'

For physicists like Pauli, it wasn't the number of particles that was so perplexing, it was the fact that no theory had predicted their existence. Most seemed to play little or no role in the processes that make stars, planets or people. Dozens of superfluous fundamental particles also appeared to make a mockery of Occam's razor.

Fortunately, one of the most influential, but least known, scientists of the twentieth century discovered a simple path out of the particle zoo.

## A fearful symmetry

Emmy Noether (1882–1935) was born in Erlangen, Germany, on 6 March 1882, daughter of the mathematician Max Noether and Ida Amalia Kaufmann. She attended school in her hometown studying

* A transparent box filled with a vapour that condenses around particle tracks making them visible.

German, English, French and arithmetic, the expectation being that she would become a language teacher, one of the few professions open to an educated woman. Emmy, however, had no interest in teaching languages and was determined instead to pursue a career in mathematics. This, she realised, would be a challenge as women could not enrol at the local University of Erlangen. However, as her father taught there, she was allowed to sit in on classes as an observer. This enabled her to pass the university entrance examination in 1903 aged twenty-one, opening the doors to the prestigious University of Göttingen.

FIGURE 35: Emmy Noether.

In 1903, the University of Göttingen was the centre of the mathematical universe. Emmy attended classes by mathematical giants such as David Hilbert, Hermann Minkowski and Felix Klein. But once again, she was not permitted to formally enrol. After one semester, she became ill and returned to Erlangen where the rules barring female enrolment had been relaxed.

Back in Erlangen and under the direction of her father's friend and colleague Paul Gordan, Emmy managed to complete her degree and went on to complete a PhD, achieving the highest honours and becoming only the second woman in Germany to obtain a PhD in mathematics. She was then allowed to teach at the Erlangen Mathematics Institute, but only as an unpaid tutor rather than as a formally recognised member of staff. It was in Erlangen that her interests shifted to some of the most pressing problems in early twentieth-century mathematics: abstract algebra. This is a method in which entire mathematical operations, rather than just numbers or symbols, are manipulated. She published several ground-breaking papers, bringing her to the attention of her former teachers in Göttingen.

These were heady times at Göttingen University. Einstein had just published his general theory of relativity papers and much of the mathematics faculty was captivated by the new theory and the challenge of unravelling its implications. Hilbert invited Einstein to lecture at Göttingen in June to July 1915. Einstein's lectures convinced Hilbert of the truth of general relativity, but the two scientists also uncovered a problem as the theory appeared to violate one of the fundamental principles of science, energy conservation – the principle that energy can neither be created nor destroyed. Hilbert thought he knew a woman who could help.

In 1915, Hilbert and Klein invited Noether to return to Göttingen. She accepted, but Hilbert's efforts to secure her a position, although supported by the science and mathematics faculties, were consistently blocked by the humanities professors who were unable to stomach the idea of a female professor. Hilbert's exasperation with their intransigence led to a famous outburst when he exclaimed that 'I do not see that the sex of the candidate is an argument against her admission . . . After all, we are a university, not a bath house!' Nevertheless, nineteenth-century prejudices prevailed so Emmy's classes had to be listed under Hilbert's name but with 'help from Fräulein Doctor Emmy Noether'. She remained unpaid.

Noether's teaching style was unconventional. She was famously unkempt, described as having a 'washerwoman' appearance with her hair breaking free of its pins as she enthusiastically led her students through the deeper recesses of mathematical theory. She also adopted Aristotle's style of peripatetic* teaching, often conducting her classes while walking through the countryside. Her enthusiasm, cheerfulness and mathematical insight secured her a loyal following among a group of students who became known as 'the Noether boys'. One of her colleagues, the mathematician Hermann Weyl (1885–1955), described Emmy as 'warm, like a loaf of bread'.[4] Weyl was appointed professor while Emmy remained an unpaid tutor. He declared that he was 'ashamed to occupy such a preferred position beside Emmy, whom I know to be my superior'.

David Hilbert enlisted Noether's skills to unravel the troubling issue of the apparent lack of conservation of energy in Einstein's general theory of relativity. As well as energy, theoretical physics includes several other fundamental conservation laws including the conservation of momentum, angular momentum and electrical charge. None are proved in classical physics; they were merely assumed to be true because no exceptions had been found. Like the constancy of the speed of light, they appear to be among the fundamental building blocks of our universe. When Emmy Noether applied her mathematical insight to the energy conservation problem in general relativity, she uncovered a deeper and wider relationship  between the conservation laws and the fundamental laws of physics. This relationship, often described as 'the most beautiful idea in physics', is known as Noether's Theorem.

The theorem is based on the notion of symmetry. It states that whenever there is a symmetry of the laws of physics then there is a corresponding conservation law.

---

* We don't know that Aristotle took his students on strolls as the term refers to the *peripatoi* or covered walkways of his Lyceum school. But I'm sure he must have done.

FIGURE 36: Spherical, bilateral, five-fold symmetry and asymmetry.

The first three pictures* above (top row, and bottom left) demonstrate different kinds of symmetry. The squid egg is (roughly) spherical. It can be rotated by any angle in any plane and it will look (pretty much) the same. The cuttlefish is bilaterally symmetrical so it can be divided along a single plane into two halves that are mirror images of each other. The sea urchin exhibits five-fold symmetry, whereas the final picture of coral on the right lacks any symmetry. Symmetrical objects are simpler than their asymmetric counterparts as they can be described with fewer parameters. For example, once you know one half of a butterfly, then you also know the other. In contrast, there's no way to reconstruct the whole of an asymmetrical object, such as coral, from any of its parts. Discovery of symmetries is thereby recognition of an underlying simplicity.

* All from the island of Sulawesi (Alfred Russel Wallace's Celebes) or its surrounding waters.

Noether's symmetries were more subtle and we can see how they work by imagining that you are admiring the juggling skills of your friend Alice in a local piazza. She possesses rotational symmetry if it doesn't matter whether she juggles while facing north, east, south or west. If there is no difference whether Alice walks a hundred yards or so in any one direction, then she can be said to also possess translation symmetry. If it doesn't matter whether she juggles today, tomorrow or any other time, then Alice is time symmetric. If her juggling balls were electrically charged, then charge symmetry would mean that Alice wouldn't care if she switched from positively charged to negatively charged balls.

In essence, these symmetries are an extension of Einstein's reasoning that the laws of physics have to be the same for all observers. Emmy Noether proved that whenever symmetries are present a corresponding conservation law applies. So, time symmetry implies energy conservation, translational symmetry implies conservation of momentum, and Newton's third law, that every action has an equal and opposite reaction, is a consequence of rotational symmetry.

Noether applied her new theorem to the problem that had troubled Einstein and Hilbert and showed that it was no longer a problem if the symmetries were accounted for. On reading her paper, Einstein wrote to Hilbert that: 'Yesterday I received from Miss Noether a very interesting paper on invariant forms. I am impressed that one can comprehend these matters from so general a viewpoint.'[5] Noether's breakthrough theorem, together with her advances in abstract algebra, brought her from obscurity into mathematical aristocracy. She was invited to present talks at prestigious conferences throughout Europe and became a member of distinguished learned societies. Despite all of this she still didn't have a job.

In 1928 she accepted an invitation to take up a post as visiting professor at the University of Moscow. Her return to Germany coincided with the rise to power of the Nazis. As a Jew and communist sympathiser Noether was expelled from the University of Göttingen in 1933 but she carried on teaching in her own home, cheerfully welcoming even brown-shirted students into her impromptu classes.

The rise of Nazism led to the flight of both Albert Einstein and Hermann Weyl from Germany. They were both offered, and accepted, positions at the prestigious Institute for Advanced Study in Princeton in the United States. Weyl lobbied hard, but unsuccessfully, for Noether to be admitted to Princeton. She was eventually offered, and accepted, a position at nearby Bryn Mawr Women's College in Pennsylvania, where she moved late in 1933.

Bryn Mawr provided Emmy with a haven away from the horrors unfolding in Europe; she could continue the work she loved, teaching and supervising research students. Sadly, her happiness was short-lived. In 1935, she was discovered to have a tumour in her pelvis and she died following surgery, aged fifty-three. On Sunday 5 May 1935, the *New York Times* published a 'Letter to the Editor', signed by Albert Einstein, entitled 'Professor Einstein Writes in Appreciation of a Fellow-Mathematician'. In fact, although Einstein signed the letter, it was composed by Hermann Weyl who wrote that 'In the judgment of the most competent living mathematicians, Fräulein Noether was the most significant creative mathematical genius thus far produced since the higher education of women began.' Many mathematicians and physicists would today drop the gender qualification. Weyl read the eulogy at her funeral saying that: 'You were not of clay, harmoniously shaped by God's artistic hand, but a piece of primordial human rock into which he breathed creative genius.'

## *Taming the particle zoo*

After her death, Hermann Weyl extended Noether's insights to develop a revolutionary approach to particle physics known as gauge theory. The theory provided a radical simplification of the particle zoo and is fundamental to modern particle physics.

Weyl's starting point was his insistence that the laws of physics should not only be independent of whether particles are here or there, rotating or not rotating, as Noether had argued, but must also be independent of how we label or categorise them. This may remind you of William of Occam's nominalist insistence that words for abstract notions, such as fatherhood, refer to mental, rather than existing, entities and thereby should not be used in science. Gauge theory takes a similar nominalist approach to particle physics[6] to find laws that are blind to how we label or describe particles or forces. Laws that possess this symmetry are described as gauge invariant.

When Hermann Weyl and his colleagues looked for gauge invariant rules for electrically charged particles, they obtained Maxwell's laws of electromagnetism. This was a remarkable result as it demonstrated that these hugely important laws, which had inspired Einstein's discovery of special relativity, were a reflection of a deeper symmetry, and thereby simplicity, in nature. Gauge theory also predicted the existence of a neutral massless particle capable of transferring forces between charged particles. Imagine Alice and Bob tossing a basketball to each other while standing, on skates, on an ice rink. When Alice tosses the ball forward then, according to Newton's third law, she is propelled backwards by her toss. When Bob catches the ball, the impact pushes him away from Alice. The net effect is that, without touching, the skaters are deflected. Weyl and his colleagues realised that this is exactly what happens when one electron meets another. A photon, like the basketball, is passed between them and causes them to veer away from each other: like-charged electrons repel.

Just as Maxwell's equations unified electricity and magnetism, so gauge theory uncovered previously hidden symmetries that unified electromagnetism with one of the two forces that hold atomic nuclei together, the weak nuclear force. Once again, entities believed to be different were discovered to be the same and the world became simpler. When applied to the strong nuclear force responsible for holding atomic nuclei together, gauge theory delivered quantum electrodynamics and the recognition that the protons and neutrons that make up atomic nuclei are actually composed of triplets of more fundamental particles called *quarks* that come in six different flavours: up, down, charm, strange, top and bottom. For example, protons consist of two up and one down quark, whereas neutrons are composed of two down and one up quark.

Gauge theory eventually resolved the particle zoo into the

Standard Model of particle physics. This envisages three generations of matter particles (Figure 37). Generations I, II and III differ only in their mass. For example, the *muon* and the *tau* are more massive versions of the electron. Then there's the single generation of the force-carrying *bosons*, including the *photon* and the rather strange *Higgs boson*. The Higgs is something of an outlier but is essential for our existence as its interaction with other particles endows them with mass. In fact it is their differing degree of interaction with the Higgs boson that endows the generation II and III particles with more mass than their generation I counterparts. If the Higgs did not exist, they would all be identical.

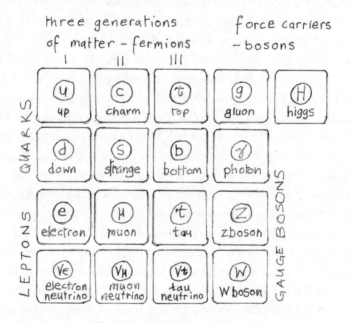

FIGURE 37: The Standard Model of particle physics.

The Standard Model is another beautifully elegant and simple scientific theory. However, most physicists believe in an even simpler world. Grand Unified Theories (GUTs) envisage that, at high energies, the electroweak and electrostrong forces are unified and *leptons* (mostly electrons and neutrinos) and quarks will be revealed to be different aspects of the same particles.

Imagine an ice rink full of skaters each pirouetting very fast on a frozen lake with a strong northerly wind blowing over its surface. When the skaters are rotating at high speed, equivalent to high temperature or energy, then they are identical and rotationally symmetrical. However, as they slow down and lose energy and the wind starts to bite, the skaters will eventually come to a halt with their shoulders aligned with the wind direction, the orientation that minimises air-resistance. Once stationary, then around half will be facing east and the other half west. Their rotational symmetry will be broken and what was a single type of skater will now be a mixture of two kinds of skater. GUTs propose that, just after the Big Bang when the universe cooled, there was a similar symmetry-breaking event that froze leptons and quarks out of a single quark-lepton particle. The particles look different but they are united through an underlying simplicity that they recover at very high energies.

## How simple does it get?

We are now close to the end of our journey following the tracks of Occam's razor from the medieval to the modern world. So far, we have taken the razor's value on trust. Yet, that trust has not been misplaced. Great scientists such as Copernicus, Kepler, Galileo, Boyle, Newton, Darwin, Wallace, Mendel, Einstein, Noether, Weyl and many others put their trust in simplicity and had that trust rewarded with astonishing new insights, advances and a simpler universe than previously imagined. But the razor does not guarantee this.

As I have repeatedly emphasised, the universe can be as complex as it likes and the razor will always be useful. Its 'nothing beyond necessity' proviso gives licence to add whatever level of complexity you need so long as you don't go beyond its remit. Nonetheless, solutions discovered by the great scientists have invariably revealed a simpler world. The same laws need not have applied in the heavens as on earth. Electricity and magnetism could have been entirely different forces; and light might have had nothing to do with either of them. Simplicity was not a given but was discovered or revealed.

The razor does not guarantee that the world will be simple, but it has nearly always turned out that way. Why?

In 1960, Eugene Wigner (1902–95), the Hungarian-born theoretical physicist and Nobel Laureate, published an influential paper entitled 'The unreasonable effectiveness of mathematics in the natural sciences',[7] arguing that the extraordinary ability of mathematics to make sense of the world is a puzzle. An analogous case can be made for the search for simplicity in science that might be called the unreasonable effectiveness of Occam's razor. In fact, Wigner makes a similar point in his essay when he argues that mathematicians seek always to discover theorems that 'appeal to our aesthetic sense both as operations and also in their results of great generality and simplicity'. Wigner's puzzlement over the 'extraordinary effectiveness of mathematics' is, then, also a reflection on the extraordinary effectiveness of simplicity. Why does it work so well?

Wigner could not explain why mathematics works so well but he concluded that 'the appropriateness of the language of mathematics for the formulation of the laws of physics is a wonderful gift which we neither understand nor deserve'. As we have discovered, mathematics is also the tool by which we discover simplicity; so the real source of Wigner's 'wonderful gift' is, I believe, simplicity.

In the final chapter of this book we will explore an extraordinary theory that may provide a reason for the 'unreasonable effectiveness' of Occam's razor. Before we get there we first need to take a closer look at how Occam's razor works and for that we must slip back a few centuries to meet a Protestant minister with an interest in gaming.

# 18

## Opening Up the Razor

The maxim which inspires all scientific philosophising . . .
[is] 'Occam's razor': Entities are not to be multiplied beyond
necessity.

Bertrand Russell, 1914[1]

In April 1761, thirty-four years after the death of Isaac Newton
and 118 years before the birth of Albert Einstein, Richard Price
(1723–91), the non-conformist Protestant minister, moral philosopher and mathematician, examined the unpublished papers of his
recently deceased friend and mathematician Thomas Bayes (1702–
61). Bayes had been a modestly successful scientist. Thirty years
earlier, he had rushed to the defence of Isaac Newton's mathematical method known as calculus, which had been attacked in an article
addressed to 'an infidel mathematician' by the Irish philosopher
and Roman Catholic bishop George Berkeley who feared that
Newton's mechanistic science would undermine religious faith. In
response, in his 'An Introduction to the Fluxions' written in 1736,
Bayes not only provided a defence of Newton but also attacked
Berkeley's motivation, arguing that 'it was highly wrong to bring
religion in at all into the controversy'. Despite being a Presbyterian
minister, Bayes went on to insist that 'I shall now consider my
subject as stripped of all relation it has to religion and merely as a
matter of human science'. The separation of at least the physical
sciences from religion that William of Occam had initiated four
centuries earlier was, by this time, nearly complete.

Among Bayes's papers was one that Price found both intriguing and puzzling. It was entitled 'An Essay toward solving a Problem in the Doctrine of Chances'. Chance, or probability, was a hot topic in the eighteenth century as the insurance business in England and Scotland had made a fortune on its ability to put a price on the risk of death, shipwreck, damage, illness, injury or any misfortune. Several of Price's family were actuaries and, ten years later, he would write his own book on statistical approaches to actuarial calculations. However, in 1761, he had never seen anything like the statistics described in Bayes's paper.

Thomas Bayes is one of the most elusive heroes of our story. We hardly know him any better than we do William of Occam. There is a widely used portrait of a rather stern dark-haired man in clerical gown and collar that is often claimed to be of Bayes, yet the attribution is, at best, doubtful.[2] He was born in 1702, probably in Hertfordshire, the son of another non-conformist minister, Joshua Bayes. After studying theology and logic at Edinburgh University he went on to become a minister at Mount Sion Chapel in Tunbridge Wells, Kent. This spa town had become a popular resort during the Restoration after the king, Charles II, and his queen visited to 'take the waters' in 1663. However, the town had thereafter earned something of a salacious reputation and, in his *The Debt to Pleasure*[3] published in 1685, John Wilmot, Earl of Rochester had described it as 'The Rendevouz of Fooles, Buffoones, and Praters/ Cuckolds, Whores, Citizens, their Wives and Daughters'.

The Reverend Thomas Bayes was not a particularly popular preacher in the 'Rendevous of Fooles', but he did establish a reputation as a scholar and was even invited to demonstrate the melting of ice to 'Three Natives of the East Indies' who visited in 1740. Probably because of his defence of Newton's calculus, Bayes was elected to the Royal Society in 1742, but published no further mathematical work before his death in 1761. Bayes's paper on probability was therefore a complete surprise to his friend Richard Price. He arranged for the paper to be read at a meeting of the Royal Society two years after Bayes's death, and then published.

## The probability razor

It seems likely that Bayes first became interested in probability after reading *A Treatise of Human Nature* by the Scottish philosopher David Hume. Hume posed what has become known as the 'induction problem' as a criticism of the dominant scientific method since the Enlightenment. As mentioned in Chapter 10, induction had been pioneered by Francis Bacon as a means of reaching scientifically valid conclusions from several observations. For example, from the observation that the sun has risen for every morning in human history, we may use the method of induction to claim that it always rises. Hume pointed out that this is not based on any solid reasoning. The propositions that 'the Sun has always risen in the morning and will rise tomorrow' is no more proven than the proposition 'the Sun has always risen in the morning and will not rise tomorrow'. Both are compatible with all the existing evidence and cannot be distinguished either on logical or on empirical grounds. Hume insisted that this method of inductive reasoning only delivers probability, not certainty.

Bayes accepted Hume's argument that induction cannot deliver certainty but he was sure that it delivers probabilities that are nevertheless useful. He set out to provide his intuition within a solid mathematical framework. As a church minister he was probably involved in fundraising events, such as tombolas, raffles or lotteries, so in his paper he starts his argument by asking us to 'imagine a person present at the drawing of a lottery, who knows nothing of its scheme or of the proportion of Blanks to Prizes in it'. At this point it will be easier to appreciate the role of Occam's razor in Bayesian statistics if we substitute dice for the tombola. We will imagine that Bayes's friend Mr Price owns two dice. The first is a conventional simple six-sided dice and the second is an unconventional and more complex sixty-sided dice. We will further imagine that Mr Price persuades the reverend to play a game in which, behind a screen, he throws just one of the two dice and calls out its number. He then asks the Reverend Bayes to guess which of the two dice he has thrown.

FIGURE 38: Dice.

The reverend's first intuition might be that Mr Price is equally likely to throw either dice. In terms of the statistics described in Bayes's posthumously published paper, he would then assign a *prior probability*, the probability before Mr Price throws the dice, of one half, 0.5, to the six-sided dice hypothesis and the same to the sixty-sided dice hypothesis. Let us say that the first number that Price calls out is 29. Bayes would surely call out 'it's your sixty-sided dice' and Price would nod his assent. Yet, Bayes's mathematical mind would also probably execute a simple calculation according to the principles that he had described in his paper. For the sixty-sided dice hypothesis, he would multiply the prior of 0.5 with a value called the *likelihood*, which is the likelihood that the sixty-sided dice would throw a 29. Since there are sixty possible numbers that the dice could have landed on, then each, including the number 29, has a likelihood of 1/60 or 0.016. Multiplying this value by the prior of 0.5, Bayes would obtain a *posterior probability* (the probability after the data has arrived) of 0.008 for the sixty-sided dice hypothesis.*

Bayes would also perform the same calculation for the six-sided dice hypothesis, multiplying its prior probability of 0.5 by its likelihood of throwing a number 29. This is of course 0 since none of

---

* Bayes's theorem for calculating the posterior probability is more correctly written as the prior probability times the likelihood divided by the probability of the observation, irrespective of the theory. I have left out the division step as it is only there to normalise the value of the posterior probabilities so that they always add up to one. We have assumed that both dice hypotheses are equally likely so the division step is not necessary in this case.

its six sides shows a 29. Multiplying any number by 0 gives 0, so the posterior probability of the six-sided dice throwing 29 is 0. Comparing the two posterior probabilities, he would divide the posterior probability of 0.008 for the sixty-sided dice by the posterior probability of 0 for the six-sided dice. Dividing any number by 0 yields infinity so the relative probability of the sixty-sided dice having thrown 29, compared to the six-sided dice, is infinity. So it is infinitely more likely that Price has thrown the sixty-sided dice. One up for Bayes.

It might seem that Bayes's theorem is making a big deal of a simple intuition but the game becomes more interesting in the second round when Price once again selects, in secret, one of the two dice. He throws it and calls out the number 5. Now the situation is uncertain as 5 could have been thrown by either dice. Are they both equally likely? The Reverend Bayes thought not and devised his statistical methods to deal with precisely this kind of inductive problem where two, or several, or even an infinite number of hypotheses or models fit the data. How do you choose between them?

The key factor in Bayesian statistics is the Bayesian likelihood, which, as first pointed out by the statistician Harold Jeffreys in his textbook on probability published in 1989[4] and then further elaborated by many succeeding Bayesian statisticians,[5] automatically incorporates Occam's razor by favouring simple theories and punishing complex ones. We can see this if we repeat the Bayesian transformation of prior to posterior probabilities for the second round of the dice-throwing game. Bayes would, once again, assign a prior probability of 0.5 for both dice hypotheses. The likelihood of the sixty-sided dice throwing a 5 is the same as that of it throwing a 29, 1/60 or 0.016. When this value is multiplied by the prior then Bayes once again obtains a posterior probability of 0.008.

However, when this operation is repeated for the six-sided dice hypothesis then its likelihood of throwing a 5 is much higher, 1/6 or 0.16. This is of course because the six-sided dice is simpler in the sense that there are fewer numbers that it can possibly throw.

Bayes multiplies the six-sided dice's prior probability of 0.5 by 0.16 to obtain its posterior probability of 0.08. This is ten times the posterior probability for the more complex sixty-sided dice. So it is ten times more likely that the number 5 has been thrown by the six-sided rather than the sixty-sided dice. On the grounds of his innovative statistics, Bayes would call out 'It's the six-sided dice' and, in this case, he wins again.

Likelihood provides Bayesian statistics with its own built-in razor that automatically favours simpler hypotheses because they have a higher probability of generating the data. Another way of visualising this is to consider the parameter space, which is the range of values that is possible for each model or hypothesis or, equally, the range of observations that each might generate. Glance at the spiral of numbers in Figure 39. Those that could be thrown by a six-sided dice, its parameter space, are in the small central region. The larger blob represents the parameter space of a sixty-sided die, whereas the surrounding space is filled with numbers that cannot be reached by either dice, stretching out to infinity. Note that the space of the sixty-sided dice includes the smaller parameter space reachable by the simple dice. The number 5 (enclosed by the smallest blob) is within both spaces as it can be thrown by either dice. It might also have been thrown by a seventy-sided dice, if Mr Price possessed one, or an eighty-sided dice or indeed an infinite number of possible dice that could have generated the same observation. This is the central problem of science that we have encountered throughout this book: model selection. When you have a profusion of possible models that account for any phenomenon, how do you choose between them? The essence of the Bayesian razor is that it opts for the theory, hypothesis or model for which the space of the data (the number 5 in the above example) occupies the largest fraction of the model's parameter space (the six-sided dice) and is thereby most likely to have generated the data. This is invariably the simplest model: Occam's razor.

The Bayesian Occam's razor is science's means of dealing with a multiplicity of models that nonetheless fit the data. Consider Newton's law that states that 'For every action there is an equal

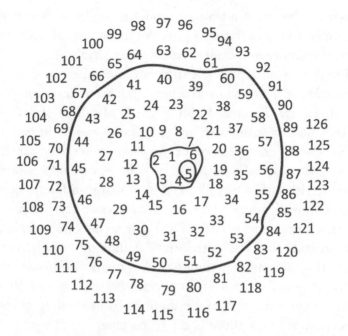

FIGURE 39: The parameter space of a sixty-sided dice.

and opposite reaction.' So, when you kick a football, the force of your boot (the action) on the ball is met with the force of the ball (the reaction) on your toe. This simple law is compatible with every kick taken in every game of football. However, there is another law that is equally compatible with all the data: 'For every action there is an equal and opposite reaction plus a small invisible demon who is also pushing on the ball to press it against your boot.' Then there is a third hypothesis with two demons and the fourth with maybe two demons and an angel providing different components of the ball's reaction to the force of your boot, and so on up to and including an infinite number of hypotheses or models.

This example is trivial, but not entirely. The aether, Ptolemy's epicycles, phlogiston, the vital principle, Henry More's 'knowing principle', the concept of a divine creator, magnetism and electricity, space and time, gravity and acceleration, quantities of energy smaller than can be measured, all of these are complex ways of turning the world's wheels. None can be eliminated on the basis of logic alone but, science insists, if a simpler model is available then it should be adopted. Bayesian statistics provides the statistical underpinning of that preference and underwrites Occam's razor.

All of the revolutionary scientific advances made by Copernicus, Newton, Mendel, Darwin and others – what the American philosopher of science Thomas Kuhn termed 'paradigm shifts' – have involved the dispensing of a more complex model in favour of a simpler one. Their preference for simple models came from mystical, theological, aesthetic principles or simple intuition. However, although Occam's razor has many different manifestations and justifications,[6] I believe the Bayesian razor expresses its essence, as applied to science. The razor favours the simple theories not because they are more beautiful, though they often are; nor because they are easier to understand, though they usually will be; nor that they make fewer assumptions, though they usually do; or because they make tighter predictions, though they always do: but because they are more likely to be true.

Nevertheless, it is important to remember that our preference for simple solutions is mostly a modern development. Prior to William of Occam, the standard response to a problem was to throw additional entities at it. William of Occam was the first to insist on drilling down to the simplest solutions, a principle that has since become the bedrock of science and the hallmark of modernity.

## The simple truth?

The Bayesian form of Occam's razor also provides an enlightening insight into the puzzle of why scientists from Copernicus to Brahe, Galileo or Newton, were so convinced that the earth orbits the sun, despite having no compelling evidence. That the giants of early modern science were convinced by heliocentricity despite lack of confirmatory evidence has been extensively cited by historians and philosophers of science, such as Thomas Kuhn[7] or Arthur Koestler,[8] as evidence that, contrary to the claims of scientists, science is not driven primarily by reason, but by irrational, personal or cultural bias. For example, Kuhn writes that: 'Judged on purely practical grounds, Copernicus' new planetary system was a failure; it was neither more accurate nor significantly simpler than its Ptolemaic predecessors.' Similarly, Koestler claimed that both Ptolemaic and Copernican models included around thirty to eighty cycles or epicycles, depending on how they are counted. On these grounds, Kuhn and Koestler argue that the simplicity criterion on which these great scientists relied was bogus.

Twentieth-century postmodernists and relativist philosophers and historians have eagerly seized on this claim to support their own assertions that science has no stronger claim on objective truth than any other system of thought. For example, the philosopher of science Paul Feyerabend (1924–94) has argued that: 'To those who look at the rich material provided by history [of science] . . . it will become clear that there is only one principle that can be defended under all circumstances and in all stages of human development. It is the principle: anything goes.'[9] According to postmodernists, science merely takes its place alongside other belief systems such as religion, mysticism, witchcraft, folk beliefs, astrology, homeopathy or the paranormal. Each has, they claim, its own truths and none can claim any monopoly on the truth. Feyerabend went on to argue that, in public education, science should not have any privileged status in school classes over mysticism, magic or religion.

William of Occam would surely have disagreed. He insisted that there was a stark difference between science and religion, as science is based on reason, religion is based on faith. Yet the postmodernists disagree. Many of their arguments are heavily influenced by the Austrian-British philosopher Ludwig Wittgenstein (1889–1951). Although trained as an engineer, Wittgenstein became fascinated with mathematics and then philosophy under Bertrand Russell at Cambridge who, in 1903, had written *Principles of Mathematics*, arguing that mathematics and logic are identical. In 1921, Wittgenstein published his hugely influential *Tractatus Logico-Philosophicus* in which he examined the relationship between language and reality and, at this stage of his career, appears to accept (philosophers still argue about the meaning of much of

Wittgenstein's philosophical statements) that science can make statements about the world that are verifiably true. Thirty years later, in his *Philosophical Investigations*, Wittgenstein seems to abandon the quest to discover how language represents the world and instead argues that there are only different ways of using language, or 'language games', whose meaning is derived solely from their use. This argument seems to have much in common with William of Occam's nominalist insistence that words refer to ideas in our heads, rather than universals or essences that exist in the world. Seven centuries earlier Occam had used the analogy of the owner of an inn hanging up an empty barrel hoop above his door to signify that wine is to be found inside.[10] The hoop has no direct relationship to a mug of wine but instead represents a convention that is recognised by other users of the convention, such as wine drinkers, to help navigate their world. All language, Occam argued, similarly gains its meaning from the utility of its terms to its users.

Wittgenstein, however, went further than Occam to argue that each language game is incommensurate, in the Aristotelian sense of a circle being incommensurate with a straight line because it belongs to a different category of being. Occam debunked that notion in the fourteenth century by uncoiling his metaphorical rope and demonstrating it could be a circle or a straight line but incommensurability and category errors have stubbornly clung on in philosophical circles. A favourite example from the British philosopher Gilbert Ryle (1900–76) imagines a visitor to Oxford who visits its many libraries and colleges but then asks, 'But where is the University?' The visitor's error is to assume that the university is a member of the category of material objects, such as a college building, rather than an institution which exists causally only in the minds of the students, staff and visitors. Similarly, Kuhn argued that: 'The . . . scientific tradition that emerges from a scientific revolution is not only incompatible but often actually incommensurable with that which has gone before.'[11] Similarly, the American postmodernist philosopher Richard Rorty insisted that 'I do not think there are . . . any truths independent of Language.'[12]

Despite their anti-scientific stance, I do have sympathy with many of the points made by the postmodernists and relativists, particularly their opposition to the notion that Western cultural values, by which is usually meant the values of wealthy, white, highly educated Westerners, are universal. They correctly point out that there are no objective grounds for assigning higher values to, for example, Shakespeare's *Hamlet* as compared to Marvel's Spider Man stories or Ashanti folk tales about Anansi the spider. Science is also a product of language and culture. As the founder of quantum mechanics Niels Bohr observed: 'We are trapped by language to such a degree that every attempt to formulate insight is a play on words.'[13] Yet this is where I part ways with the postmodernists. Their relativistic insights cannot be transferred to science because, unlike culture, scientific laws are written in the universal language of mathematics. The relationship of the square on the hypotenuse to the squares on the other two sides of a triangle has been known for millennia by the ancient Babylonians and Egyptians and people all around the world irrespective of language and culture. It is not relative to anything.

This is why William of Occam's freeing of mathematics from the shackles of incommensurability and metabasis prohibitions (Chapter 5) was so important. Centuries later, it allowed Galileo and Newton to nail terrestrial and heavenly motions to the same set of simple numerical rules that would have been comprehensible to an ancient Babylonian tax collector, a Mayan astrologer or an African trader. Mathematics drills reason down to its simplest possible set of rules and thereby lifts science from being just another game into becoming a universal language.

There is, however, another postmodernist insight that is both true and crucial to the role of Occam's razor, though it does not lead down the path taken by the postmodernists. Truth really is, as they argue, unknowable. This is something that is shocking even to scientists who are generally taught that science is an inexorable march towards the truth.

Imagine that science were able one day to attain the blissful state of knowing everything, i.e. 'the truth'. How would we know? Knowledge of ultimate truth presupposes some means of peeking behind the curtain of evidence provided by our senses or scientific instruments to see the 'real' world rather than the one viewed through our senses or scientific instruments. It supposes that there is some knowable, complete and perfect world, a world of idealised Platonic forms, the very view of the world Occam disproved so many centuries ago. If like Occam we reject this view of the world we have to rely instead on our sensory inputs and a potentially infinite variety of models of the cosmos that could fit that data and explain our place in it.

Yet that does not mean, as the postmodernists argue, that all models are equal. When drawing up a horoscope, today's astrologers do not consult accounts of the moodiness of the god Mars or the lustful habits of Jupiter. They instead turn to planetary tables based on Kepler's simple model of the solar system. Believers in the paranormal organise their meetings by phone and email, not telepathy; and, if the meetings are overseas, they fly by plane, not levitation. Science may be a language game or model but, unlike the vast majority of models, from alchemy to feng shui, homeopathy and the indecipherable postmodernist tracts that dismiss science, its models actually work because they are simple and thereby deliver accurate predictions.

## Science is simplicity

Nearly all science, indeed nearly all our knowledge of the world, is based on Bayesian reasoning applied to inductions. As the post-modernists insist, the evidence of a thousand observations of the sun rising does not provide us with certainty, but it does provide us with the high likelihood that the simplest hypothesis – that it will also rise tomorrow – will turn out to be true. Probability, rather than certainty, is sufficient for science and is the core of modern science. Alchemists experiment, astrologers calculate, but neither they, nor a thousand other mystics, philosophers or priests, insist upon accepting only the simplest solutions because they are also the most likely.

Of course, science is not only about simplicity. Experiment, logic, mathematics, repeatability, verification and falsifiability all play vital roles. The last of these, falsifiability, was championed by the philosopher Karl Popper (1902–94) and is probably the most cited criterion for distinguishing science from pseudoscience. Yet it does not guarantee science because proving a theory false is as impossible as proving it to be correct. Any experimentalist knows that when they perform an experiment that generates a result contrary to their prediction, they will not rush to declare that their favourite theory has been falsified. Instead, they will confabulate reasons for why they observed the contrary data by adding extra complexities. We saw this process in action, in Chapter 12, when the adherents of phlogiston or caloric invented new entities, such as negative weight, rather than abandon their theories. Creationists are masters at inventing preposterous, but non-falsifiable, complications that would account for facts such as the fossil record.

The inability of data to disprove a theory is also apparent in the observation that dead theories, apparently disproved by solid evidence, are occasionally raised from the dead. For example, Lamarckian inheritance was supposed to have been disproved by observational and experimental evidence against inheritance of acquired features, such as the muscular hammer arms of blacksmiths. Yet evidence for limited inheritance of several acquired characteristics, such as dietary preference, emerged in the 1990s to revive a form of Lamarckian inheritance known today as epigenetics.[14] In the twentieth century Einstein invented a factor called the *cosmological constant* to make his theory of general relativity consistent with a static universe. He abandoned it when the universe was seen to be expanding. Yet in the twenty-first century, the cosmological constant has been revived to account for the dark energy of space itself. Similarly, as we discussed in the last chapter, nobody has ever disproven geocentricity because it's not wrong. It just doesn't work as well as the alternative. The postmodernists are right, in the final analysis, that no theory can ever be proved right or wrong. Yet that does not stop us from choosing the simplest that correctly predicts the facts. Simplicity, rather than falsifiability, lies at the heart of science.

## *The pocket razor*

Of course, we don't need to consult the Reverend Bayes to recognise, for example, that the heliocentric model provides a much simpler account of the path of the planets across the sky than the geocentric confusion of wheels within wheels. It is obvious. Our mind seems to have a natural preference for simplicity and, as argued by the cognitive psychologist Nick Chater,[15] automatically assigns higher probabilities to simpler models. But how do we recognise what is simpler? One feature that is easily assessed is the length of an explanation. Shakespeare recognised that 'brevity is the soul of wit' but it is also the signature of simplicity. Tall tales tend to be lengthy tales. The philosopher Nelson Goodman devised a rather complex test of textual simplicity,[16] but a much simpler rule of thumb that works well is what I might call Occam's pocket razor. It counts the number of significant words (excluding articles, conjunctions, etc.) required for rival explanations or models and punishes the lengthier by halving its probability for every significant extra word.

If we apply the razor to the two models of the planetary paths, then accounting for the paths by placing the sun at the centre of our model would take, say, fifty words. Accounting for the same paths by placing the earth at the centre and erecting all those epicycles in the heavens would take, at a conservative estimate, at least one hundred words. According to my pocket calculator, our pocket razor suggests that the heliocentric system is two to the power of seventy or about a million billion fold more likely than the geocentric model.

Try applying Occam's pocket razor to other disputes that we have met in this book, such as a creationist versus natural selectionist account of fossils or 'figured stones'. It is also instructive to test its edge on pseudoscience remedies such as homeopathy or crystal healing, explaining how they work against the rival explanation that they simply don't. Global warming and its likely causes provides another interesting whetstone on which you might like to sharpen your pocket razor.

Finally, I want to remind you that, in itself, Occam's razor makes no claim on the simplicity or complexity of the universe. Instead, it urges us to opt for the simplest models that can predict the data. We can call this form of the simplicity principle the weak Occam's razor. However, many scientists, particularly physicists, accept what might be called a strong Occam's razor that claims that the universe is about as simple as it can be, given our existence.

# 19

## The Simplest of All Possible Worlds?

Every action done in nature is done in the shortest way.
                                    Leonardo da Vinci, notebooks[1]

It is the summer of 1753 and in Berlin the city's hangman is burning books on the order of Frederick the Great. The words printed on the blazing pages are those of the giant of the French Enlightenment, and at that time Berlin resident, Voltaire (1694–1778). His pamphlet *Diatribe du Docteur Akakia* is a vicious lampoon of the life and work of another Frenchman, Berlin resident and the president of the Berlin Academy, Pierre Louis Moreau de Maupertuis (1698–1759). Maupertuis was born in 1698, four years before Thomas Bayes, in Saint-Malo, a port on the Brittany coast of France. He studied mathematics in Paris and in 1723 was admitted to the Académie des Sciences where he became a champion of the mechanistic laws that Isaac Newton had discovered across the Channel. In the 1730s Maupertuis engaged in a debate over whether the earth was flattened at either equator (prolate) or the poles (oblate). He used Newtonian mechanics to predict that the earth was oblate, in opposition to the opinion of the eminent French astronomer Jacques Cassini (1677–1756). In 1736, Louis XV of France appointed Maupertuis to lead an expedition to Lapland to settle the dispute. Maupertuis's measurements of the curvature of the earth in its northerly extremity demonstrated that our planet is indeed flattened near the poles. The work so impressed Frederick the Great, who had just founded the Berlin Academy, that he offered the post of its director to Maupertuis. In 1745, the Frenchman accepted.

Around this time Maupertuis became interested in an even loftier project: he thought he could prove what Aquinas had attempted five hundred years earlier: the existence of God. Light, which the medieval theologian, Robert Grosseteste, believed to be an emanation from God, provided the clue. A century earlier, the French mathematician Pierre de Fermat (1607–65) had puzzled over why light rays bend when they travel from one medium to another. This phenomenon, which we call *refraction*, is responsible for making a straight stick (or pencil) appear bent when partly submerged in water. At first glance, the phenomenon contradicts the principle of simplicity since a bent line is a kink more complex than a crooked one. However, Pierre de Fermat had already proposed that instead of minimising the number of kinks in its paths, light instead minimises the time it takes to get to where it's going. He combined this with his conjecture that light travels more slowly in water to account for the bent path by proposing that light takes the shortest path through the slower medium in order to minimise total travel time (Figure 40). In 1744, Maupertuis demonstrated a more general *principle of least action* that applied to both refraction and reflection in which he proposed that both phenomena minimise, not time, but time multiplied by energy, what is called 'action'.

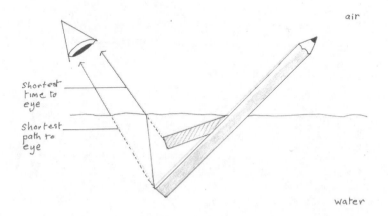

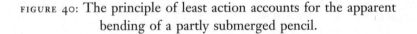

FIGURE 40: The principle of least action accounts for the apparent bending of a partly submerged pencil.

To get to grips with the principle, it will be useful to return briefly to the ancient problem of understanding the flight of an arrow after it has left the archer's bow. The problem had vexed Aristotle, prompting Jean Buridan to invent the concept of impetus as a kind of fuel that powered the arrow. Any archer knows that, to hit a distant target, the arrow must be aimed high so that its fall terminates at the target. Each trajectory is uniquely determined by the angle and speed of flight from the bow. An experienced bowman will know the right path, but how does the arrow *know* which path, out of many possible alternatives, it must take once it has left the bow (Figure 41). One answer is that Newton's laws predict the correct, and unique, parabolic trajectory from bow to target. Nevertheless, the calculations included a value for that circular concept of 'force'. Maupertuis discovered that he could obtain the same trajectory if he dispensed with force and instead assumed that, for the entire journey, the motion of the arrow minimises its *action*.

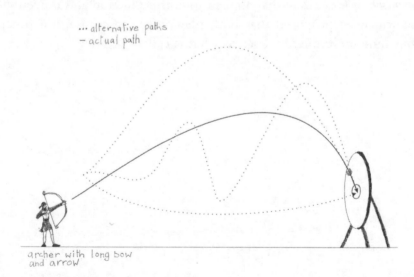

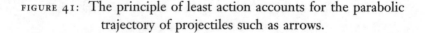

FIGURE 41: The principle of least action accounts for the parabolic trajectory of projectiles such as arrows.

Let's imagine that we replace the archer's bow with a fuel-powered rotor mounted on the shaft of the arrow that powers its flight from archer to target. We will further imagine that the arrow's rotor speed and direction is controlled by a tiny on-board computer that adjusts its speed and trajectory to a path that minimises fuel consumption for the entire journey. Remarkably, the path taken by our computerised arrow will be the same as a real arrow because both minimise action.

The principle of least action states that the motion of any body, such as an arrow in flight, will take a path that minimises its total action. This is the sum, for each point on its journey, of its kinetic energy due to its motion minus its potential energy, due to its position in an energy field such as the earth's gravitational field. For powered locomotion, the action will roughly correspond to the quantity of fuel energy consumed during that journey. The principle of least action states that, for natural motion, this action is minimised. It governs the path of arrows, rockets, planets, electrons, photons or indeed any kind of particle and even waves.

Moreover, and most remarkably, scientists have since discovered that many or most fundamental laws of physics can be derived from the principle of least action. For example, considering the motion of classical objects such as arrows or cannonballs, minimising action delivers precisely the trajectories predicted by Newton's three laws of motion. When applied to quantities such as energy, momentum or angular momentum, then it uncovers the classical conservation laws as well as Noether's theorem and the gauge theories of particle physics. For quantum particles, such as photons, the principle of least action delivers Richard Feynman's path integral method of calculating particle motion.* When light rays bend between a submerged stick and your eye, they do so along the path of least action. The same law ensures that stars, planets or even black holes follow their paths of least action through

* In his *Lectures on Physics* (1964), vol. 2, Chapter 19, Feynman tells us that, after his high school teacher told him about this principle, he found it '. . . absolutely fascinating'. He went on to write his PhD thesis at Princeton University on the application of the principle of least action in quantum mechanics.

gravitational fields, the same as those predicted by Einstein's theory of general relativity.

The extraordinary generality of the principle of least action and its ability to deliver so many 'fundamental' laws suggests that it is a very deep principle and one that, according to South African-born physicist Jennifer Coppersmith, shows that we inhabit a 'lazy universe'. Back in the eighteenth century Maupertuis claimed that his discovery that 'nature is thrifty in all its actions' proved the existence of God.[2] In his 'Derivation of the Laws of Motion and Rest as a result of a Metaphysical Principle', published in 1748, he argued that: 'These laws, so beautiful and so simple, are perhaps the only ones which the Creator and Organizer of things has established in matter in order to effect all the phenomena of the visible world.'

Instead of the acclaim that he had expected, intellectuals from all over Europe ridiculed his idea. To make matters worse, his claim for priority in the discovery of the principle of least action came to be disputed by several scientists, most famously the German mathematician Johann Samuel König (1712–57). König's case was championed by Voltaire. When, through Maupertuis's influence, the younger König was forced to leave the Berlin Academy, Voltaire was provoked to pen his *Diatribe du Docteur Akakia*. Frederick the Great rallied to the support of the director of his academy by ordering the book-burning. Nevertheless, Maupertuis felt humiliated. He resigned from the Berlin Academy and returned to Paris; finding little support there, he moved on to Basel in Switzerland, where in died in 1759.

History has been kinder to Maupertuis and he is generally credited with the discovery of one of the deepest principles in science, the principle of least action. Just as the speed of light is the same for all observers, Maupertuis's principle is not predicted by any more fundamental law but appears instead to be part of the bedrock of our universe. It is a *strong* Occam's razor that insists that, for the universe, *action* should not be multiplied beyond necessity.

Nevertheless, despite the principle, our universe remains highly complex with lots of *stuff* that appears to be beyond necessity. For example, neutrinos, those particles that Enrico Fermi predicted in 1931, are extraordinarily numerous yet hardly interact with any other particle so that trillions are passing harmlessly through your body every second. Would the universe not be simpler without them? Moreover, as we have discovered, although the Standard Model is relatively simple with only seventeen particles, it could be simpler. What is the point of the majority of quarks and leptons in the Gauge Groups II and III that do not contribute to ordinary matter? You have probably also heard of two other entities that appear brazenly beyond necessity: the dark matter and dark energy that make up most of our universe. Why hasn't the universe, perhaps under the influence of the principle of least action, minimised all the dark stuff?

To justify my assertion that the universe is close to being as simple as it could be, I first have to find roles for quite a lot of entities that appear to be beyond necessity. We will start our search at the site of an ancient catastrophe.

## Winter is coming

Sixty-six million years ago, in the period we know as the Late Cretaceous (100–66 million years BP) the climate and oceans were much warmer than today, spawning a diverse and abundant fauna on land and in the sea. In the oceans, the spiky foraminifera (like amoeba but with shells), which fed on photosynthetic microbes and algae, lived, died and were fossilised to become the rolling hillsides of the chalky North Downs in Surrey where the young William of Occam perhaps walked as a child. Arthropods, molluscs, worms, anemones, sponges, jellyfish, echinoderms and the squid-like belemnites and their cephalopod cousins, the spiral-shelled ammonites, grazed on the foraminifera until their shells sank to the sea floors to become, after millions of years, the figured stones that puzzled those eighteenth-century naturalists who pondered the origin of species. At the top of the oceanic food chain were the fish and marine reptiles such as the long-necked plesiosaurs and gigantic mosasaurs whose fossilised skeletons Mary Anning chiselled out of the Dorset cliffs. On land, herbivorous dinosaurs, such as the duck-billed hadrosaurids and the beaked triceratops, roamed through coniferous forests or waded through swamps buzzing with insects that pollinated the profuse flowering plants. Never far away were the roars of the big land predators, such as Tyrannosaurus rex, while flying lizards, including several pterosaurs, hunted from the air.

Yet all of these creatures were about to become extinct. A 10-kilometre-wide rock, which had been harmlessly moving through the edge of our solar system for billions of years, had been deflected from its orbit by a bend in space-time caused by the earth's mass. That bend, which we know as gravity, diverted the rock's trajectory into the kind of parabolic path that Galileo would have admired but one that intersected with our planet's thin terrestrial surface. In the last seconds before impact, the rock was accelerated to a speed of 10 kilometres per second, about twenty times faster than a speeding bullet.

Any dinosaur that happened to glance skyward in those fateful seconds would have seen a fireball, far brighter than the sun, streak across the sky. It would have been followed by an intense flash of light as the asteroid slammed into the Gulf of Mexico, vaporising in an instant about a thousand cubic miles of rock and liquefying a vast region of the earth's crust to gouge a crater 180 kilometres wide and two kilometres deep. The ejected debris formed a dense dust cloud that blocked out the sun, initiating a deadly winter that lasted for decades.

About 80 per cent of all species on our planet, including all the dinosaurs (except their cousins, the birds), were wiped out in what has been called the 'worst weekend in the history of the world'. Yet, it was not actually the worst. There have been five mass extinctions in the history of our planet and a host of minor ones. The Permian-Triassic extinction event about 250 million years earlier had been far more destructive as it had nearly sterilised the planet by wiping out a staggering 96 per cent of all known species. In fact, a series of mass extinctions have regularly interrupted the leisurely progress of natural selection on our planet.

Palaeontologists David M. Raup and J. John Sepkoski of the University of Chicago found evidence for a pattern of mass extinctions of about one every 26 million years. Since no terrestrial cycles are known with such a long periodicity, their discovery prompted a search for answers in the heavens. One of the most controversial is the theory put forward by theoretical physicists Lisa Randall and Matthew Reece, of Harvard University, Massachusetts. They reason that dark matter killed the dinosaurs.[3] Randall and Reece proposed that our solar system's rotation around the galaxy periodically brings it close to a thin disc of dark matter in the galactic plane that disturbs the orbits of comets and asteroids to send showers of rocky destruction tumbling towards the earth.

At first sight, it would seem that both neutrinos and dark matter are entities beyond necessity in our universe, particularly if you happened to be a dinosaur. To discover why neither we, nor dinosaurs, would have existed without them, we need to probe the origins of matter and particularly its living variety.

## It takes a galaxy to raise a planet

In 1915, Albert Einstein applied his theory of general relativity to the entire universe. To his surprise, he discovered that the universe it predicted was not stable: it must either contract or expand. To counter this instability and generate a static universe, Einstein added a cosmological constant, which is a kind of energy of space providing a kind of pressure against contraction. However, in 1929 the astronomer Edwin Hubble measured the velocities of galaxies and discovered, to his astonishment, that they are nearly all moving away from us as the universe is expanding. Einstein abandoned his cosmological constant, calling it the 'biggest blunder' of his life.

If the universe is expanding into the future then it must have been much smaller in the past. If we take our current universe and run the clock backwards then we can predict that, around 13.8 billion years ago, when our universe was only about a second old, all its matter was crushed into a super-hot sphere about the size of an apple filled with a kind of gas of fundamental particles. This apple-sized universe subsequently expanded in the Big Bang to provide the flash of radiation that, 13.8 billion years later, Arno Penzias and Robert Woodrow Wilson detected with their horn antenna parked on a hilltop in New Jersey. Yet Penzias and Wilson would never have been on that hilltop, nor would there have been a hilltop or New Jersey, if it wasn't for neutrinos.

The first role that neutrinos played in our existence was in helping to ignite stars. As the universe continued to expand following the Big Bang, hydrogen and small amounts of helium began to coalesce, under the influence of their gravity, to form proto-stars. Initially, these were dark objects but, as their density increased, hydrogen protons fused to form helium nuclei that ignited the stars' nuclear furnaces and released the universe's first bursts of starlight. Neutrinos are essential for this reaction. They are needed to satisfy one of those conservation laws required by Emmy Noether's theorem; in this case, the law of the conservation of leptons. The law demands that the total number of leptons (electrons, muons, tau particles and neutrinos) remains constant. This can only happen in stellar nuclear fusion through the release of massive numbers of neutrinos in starbursts. So, far from being superfluous, if neutrinos didn't exist then the universe would be dark and lifeless.

Neutrinos also played an important role in distributing elements essential to life to regions where life might emerge. The Big Bang made hydrogen and helium but no carbon, nitrogen, phosphorus or sulphur. These heavier elements essential for life were made by nucleosynthesis within the super-hot interiors of stars but could not be accessed by life while they remained locked within the hottest objects in our universe. Which is where those tiny neutrinos made another big difference.

The fate of each star depends on its size and composition. If smallish, like our sun, then when they run out of hydrogen fuel, they expand to form red giant stars before contracting to become inert white dwarf stars that, once again, lock away all their precious heavy elements essential for life. However, massive stars, more than ten times bigger than our sun, tend to go out with more of a bang than a whimper. Their gravitational attraction makes them predatory, swallowing smaller nearby stars in a gravitationally positive feedback loop until they collapse to form a neutron star. The centre of a massive neutron star may condense further to become a black hole, again locking away all their elements essential for life. However, while the neutron star is collapsing it sets off a shock wave that triggers an expansion of the outer shell of the star that initially falters, but is then reignited when a blast of neutrinos flies out of the core. It is this neutrino ignition that creates the most energetic event in our universe, a supernova explosion, such as the one that Tycho Brahe observed in 1572.

Supernovae are responsible for blasting the heavy elements such as carbon, oxygen and phosphorus, essential for life, into cool places where life can make good use of them, such as on our own planet. It is something of a cliché these days, but, as Joni Mitchell sang in 1970, we are indeed made of stardust but distributed by loads of tiny neutrinos. Far from being entities beyond necessity, the universe would be a very dull place without these almost massless neutral particles.

## A dark ecology

Dark matter, the material that may have been responsible for exterminating the dinosaurs, makes up about 27 per cent of the universe. A whopping 68 per cent of our universe is made of another mysterious entity known as dark energy. The visible sun, stars and planets account for only 5 per cent of matter-energy. Why has the universe wasted so much of its resources making such a lot of dark and apparently superfluous stuff?

In fact, far from being an entity beyond necessity, dark matter has played at least two key roles in our existence. The first was to help make galaxies. This was something of a puzzle because, as Neil Turok noted (see Introduction), the cosmic microwave background (CMB) is extremely smooth, indicating that at its birth the universe was very simple, being very smooth and rather dull. If it had stayed this way, then galaxies and stars could not have formed. However, if the CMB (Figure 2) is examined very closely and its irregularities amplified, then lumps, clumps and strings of slightly denser material are discernible. Dark matter appears to have played a key role by acting as a kind of clotting agent that helped to coalesce the diffuse gas into the lumpy clouds that became galaxies, stars, planets and eventually us.

Another role for dark matter emerged from the observation that old galaxies, like our own Milky Way, are continuing to make new stars at a rate of about one per year, mostly at their fringes. This is a puzzle as it was thought that the raw material for stars had mostly been formed in the Big Bang and, by now, would have been exhausted. Supernovae do play an important role in recycling stellar material but, when they explode, their *ejecta* are fired into space at a speed of about 1,000 kilometres per second. Remember that space is, mostly, space. So, there's hardly anything to stop supernovae remnants, with all their life-essential heavy elements, from hurtling right out of our galaxy to be lost for ever within the vast reaches of intergalactic space. If this were the fate of most supernovae ejecta then galaxies would have been stripped of their interstellar gas and dust long ago, causing the engines of star formation to stall.

The reason this has not happened was provided by an observation made by the American astronomer Vera Rubin. Born in 1928, Vera became fascinated by astronomy from about the age of ten. At the age of fourteen, she built her own telescope and, by the time she left school, she harboured a determination to become a professional astronomer. But this was the 1940s and attitudes towards women in science were no more progressive in America than those Emmy Noether experienced in Germany. When Vera applied to major in science at Swarthmore College, Pennsylvania, the admissions officer asked what career she wanted to pursue. She described her ambition to become an astronomer. The interviewer looked doubtful and asked if she had any other interests. Vera replied that she liked to paint. The interviewer responded, 'Have you ever considered a career in which you paint pictures of astronomical objects?'[4] The line became a favourite jest in her family. Whenever anyone made a mistake, someone would quip, 'Have you ever considered a career in which you paint pictures of astronomical objects?'

Vera was however undeterred by her early career advice and became one of the most eminent astronomers of her generation. In 1965, she took up a position at the prestigious Carnegie Institution of Washington where, with her colleague Kent Ford, she embarked on a project to measure the mass distribution within galaxies by measuring the speed of rotation of stars. Newton's laws insist that the gravitational force is proportional to mass; and since most of the mass of a galaxy was supposed to be concentrated at its central bulge, it was expected that the inner stars would rotate fast and the outer stars more slowly, just as the outer planets orbit the sun much slower than the inner planets.

But when Rubin and Ford measured the rotation of stars in the nearby Andromeda galaxy, they found that the distance of a star from the centre of the galaxy did not affect its speed at all. Instead, the outer stars rotated just as fast as those close to the centre. At first, Rubin didn't believe the data but she and Ford repeated the measurements for more and more stars and obtained exactly the same result. They eventually concluded that spiral galaxies contained about six times more matter than visible stars and that this dark matter was responsible for accelerating the outer stars of the Andromeda galaxy.[5] Rubin and Ford's observations have subsequently been confirmed by many more observations of distant galaxies. Each seems to be surrounded by a halo of unseen cold dark matter.

Continuing star formation is due to this invisible halo of dark matter. The halo acts as a kind of gravity shield that deflects much of the supernovae ejecta back into the galaxy. There it can condense to form new stars, such as the sun, but also rocky planets such as the earth. We are the ejecta of exploding stars; but we needed dark matter to steer its heavy elements into life-friendly places.

Having found a role for dark matter in our existence, can we also find a role for dark energy? Not yet. It remains too mysterious.

The only sign of its existence is the surprising discovery that the universe's expansion appears to be accelerating. However, given that we know nothing about the nature of dark energy it is not possible to speculate what its role is in our existence but, if I am right, it will have one.

There remain a handful of other candidates in our universe for entities beyond necessity. The generation II and III matter particles in the Standard Model differ only in their mass from generation I particles and appear to be beyond necessity as they are absent from regular matter. Roles may yet be discovered, perhaps in the synthesis of heavy elements inside stars or within supernovae or the elimination of deadly antimatter from our universe.[6] Also, for the universe to be the way it is, the laws of physics have to distinguish between particles and antiparticles, as well as between time going forward and time going backwards. Three generations of particles seem to be the minimum needed for the universe to make these distinctions.

So, despite a few loose ends, it remains likely that we inhabit a universe that is close to being as simple as it could possibly be while remaining habitable. But why? To approach this question, we must first tackle a tuning problem.

## *The incredible unlikelihood of being*

The Standard Model of particle physics includes not only the particles, such as quarks, electrons or photons and the forces that act between them, but also some very unlikely numbers. These include the masses of all the fundamental particles as well as the strength of the forces that act between them. These values are not predicted by any theory. Instead they are fitted to the data extracted from particle colliders in pretty much the same way that Ptolemy fitted his epicycles to astronomical data two thousand years ago. Their values are, as far as we can tell, arbitrary.

Remember the imaginary dice game played by our Tunbridge Wells statisticians, Bayes and Price. Imagine that the Reverend Bayes keeps all his important papers in a safe secured by a combination lock with ten tumblers, each of which has to be turned to a particular number between 0 and 60 before the door will open. However, the reverend has mislaid the combination and cannot open the lock. Mr Price visits just as the reverend is despairing of ever recovering his papers. He happens to have brought with him his sixty-sided dice and, just to distract the exasperated Bayes, he suggests throwing the dice to see if it can discover the lock's combination. Bayes is, of course, highly sceptical but lacking any alternative, he agrees to give it a go. Price throws the numbers 55, 23, 48, 5, 76, 22, 35, 59, 41, 8. Bayes turns the tumbler to this combination and is astonished to discover that it opens the lock.

Bayes's astonishment is well founded. The chances of Price randomly coming up with the correct number is 60 to the power of 10, $60^{10}$. To put it another way, the Bayesian likelihood of Price throwing this particular number is 600 million to one. Of course, Price's throws may have been sheer luck, but the Reverend Bayes would probably search for a simpler explanation, such as trickery.

The values of the fundamental constants are similarly extremely rare and unlikely numbers but whose precise value is essential for our existence. Consider, for example, the masses of the electrons, protons and neutrons that make up atoms. If we assign a value of 1 to the mass of the proton then the electron weighs in at a tiny 0.0543 per cent of the proton's mass, whereas the neutron's mass is, like the proton's, also 1. There is no reason that we know of for any of these masses, yet change them by just a fraction and we would not exist.

For example, the mass of the neutron is not quite the same as the proton's as, if we define the proton's relative mass more precisely as 1.000, then the neutron's mass comes in as 1.001, just 0.1 per cent heavier than the proton. Isn't that a bit strange? It is as if some god or law of physics demanded that their masses be the same but someone got the sums wrong by just a fraction. Maybe it doesn't matter? In fact it does matter, hugely, as this 0.1 per cent difference is responsible for the bizarre Jekyll and Hyde character of the neutron that is essential for the world being the way it is.

First the nasty Hyde character. Free neutrons are highly radio-active and rapidly decay to a proton, electron and antineutrino with a lifetime of about fifteen minutes. This is a very fast decay, around 1,600 times faster than even a highly radioactive element, such as plutonium. The instability of the neutron is because its small extra mass is just enough for it to decay into both a proton, electron, plus a nearly massless neutrino. This reaction plays a key role within the stellar nuclear fusion reaction that is responsible for the nucleosynthesis of the heavy elements essential for life. Yet neutrons also make up about 20 per cent of the mass of our bodies. If the neutrons in the atoms inside our body were this radioactive then our flesh would disintegrate in minutes.

The fact that our flesh does not fall from our bones is due to the well-behaved Jekyll character that neutrons adopt when they are locked inside atoms. Here, that tiny extra mass that was enough to allow the neutron to decay outside of an atom, is insufficient to allow the same event to take place inside an atom. This is because, for the decay to proceed, it has to have enough energy to overcome the nuclear binding forces. This requires just a fraction more energy than that provided by the 0.1 per cent extra mass. So neutrons are well behaved and stable inside atoms. Yet if the neutron was 0.1 per cent heavier than it is, just 0.2 per cent heavier than the proton, then neutrons would decay inside atoms and matter, as we know it, would be impossible.

Things would not be any better if the neutron was a bit lighter. If the neutron was just a fraction less massive than the proton, then free neutrons would be stable so that they, rather than protons, would have been the principal product of the Big Bang. Neutrons are uncharged so, unlike protons, they are incapable of attracting negatively charged electrons to form stable hydrogen atoms that form stars. A hypothetical universe dominated by stable neutrons would be without atoms, matter, stars, planets or us. That little bit of extra mass is not only required, but it needs to be in the right direction.

Physics is full of these 'coincidences' and odd values that appear to be finely tuned to the requirements of a life-friendly universe.

This was first realised by several scientists including the English and American physicists John Barrow and Frank Tipler who wrote *The Anthropic Cosmological Principle*[7] in 1986. They highlight many more parameter values and coincidences in physics that are unaccountable yet essential for our existence. If we return to our Bayesian lock analogy, it's as if a life-friendly universe, such as the one we inhabit, depended on the throwing of a vast number of multi-sided dice falling on the right number to provide the combinations to hundreds of safes. Understanding how the universe came to have these values is known as the fine-tuning problem.

Barrow and Tipler argue that these extraordinarily unlikely values can only be accounted for by the fact that we live in an *anthropic universe* in the sense that, if the fundamental constants had different values then we would not be here to lament our non-existence. The anthropic principle does not account for the unlikely values, it merely accepts their necessity for existence. It does not explain how the universe arrived at its finely tuned combination of fundamental constants nor who, or what, threw the dice.

There seems to be only a limited number of possibilities. Theists seize on these cosmic coincidences as evidence of a divine hand knowingly turning the tumblers to their precise numbers, just as William Paley had argued that a divine watchmaker must have fashioned biological complex structures, such as the eye. This is once again a god of the gaps argument that, in reality, solves nothing as it just passes the explanatory buck from the known universe to a hypothetical God.

The other solution is the multiverse theory, also known as the many-worlds or parallel-universe theory. This idea, familiar to sci-fi enthusiasts, proposes that there exists a vast, potentially infinite, number of parallel universes, each with different values of the fundamental constants. Most are sterile but a tiny fraction just happen to have their fundamental constants tuned to the unlikely values that are compatible with life. We inhabit one of the lucky universes and there's no one around in the vast majority of unlucky universes to lament their absence of atoms, stars, heavy elements, planets or intelligent life.

## The evolving cosmos

Lee Smolin was born in New York City and went on to study theoretical physics at Harvard University before engaging in a research career, first at the Institute for Advanced Study in Princeton, where both Albert Einstein and Hermann Weyl took refuge during the war. He went on to several prestigious posts before becoming one of the founding faculty members of the world-renowned Perimeter Institute in Ontario, Canada. Throughout his career, Smolin has been engaged in the search for a unified theory that would bring together all the forces and particles of physics. He is one of the architects of both string and superstring theory, which are at the forefront of attempts to unify gravity with the particles and forces of the Standard Model.

String theories (there are several of them) propose that the matter particles, quarks, electrons, protons and so on, are all expressions of very much tinier vibrating strings. However, to make the theories work, strings need to vibrate in a 26- or 10-dimensional universe. Unfortunately, with so many dimensions, the number of possible string theories explodes to an astronomically large, potentially infinite, number. Just like the Ptolemaic solar system, excess complexity allows string theories to be fine-tuned to fit almost any reality. The theory is currently incapable of making any testable prediction.

Disenchanted with the failure of string theory to connect with reality, Smolin looked for an alternative way of accounting for the fine-tuning problem. In his books *The Life of the Cosmos*,[8] published in 1999, and more recently *Time Reborn*,[9] published in 2013, Smolin argues that natural selection may account for the unlikelihood of our universe. He proposes that our universe is the product of a cosmological evolutionary process, roughly analogous to natural selection, which he calls cosmological natural selection (CNS).

## Cosmic ecology

Making natural selection work with universes requires a degree of fine-tuning in itself. Smolin must provide universes with the three vital ingredients of natural selection: self-replication, heredity and mutation. Black holes play a crucial role in each. We will start with self-replication. You will remember that black holes are the endpoint for the collapse of massive stars whose gravitational attraction has become so great that not even light can escape their pull. They are thought to lie at the centres of most galaxies, including our own, swallowing up surrounding stars to release huge quantities of energy.

One of the possible scenarios for the end of our universe is that all its matter will eventually be swallowed up by a super-massive black hole in a scenario known as the Big Crunch. Now imagine filming this grim scenario and playing the movie backwards. The first frame of such a visual record would record the Big Crunch black hole that has just swallowed up the last remnants of our universe. Since there is nothing else in the universe, no ruler with which to measure anything, the Big Crunch black hole would be a dimensionless point in a dimensionless space. However, a moment later (in the backwards-played film) the invisible super-massive black hole would spew elementary particles and energy that would, over millions of years, coalesce into atoms, stars and even inhabited planets. This played-back Big Crunch scenario would look very much like the origin of our own universe in the Big Bang.

Since the laws of physics are symmetrical in time, this time-reversed Big Crunch is just as physically feasible as the Big Bang. The symmetry of the two events leads many cosmologists to argue that what appears to be a star-hungry black hole in our universe may be the Big Bang of another universe on the other side. Conversely, the other side of the Big Bang that initiated our universe, Smolin argues, might have been the Big Crunch of its parent universe. So, according to Smolin (and also many other cosmologists), time did not begin at the Big Bang but continues backwards through our Big Bang to the Big Crunch death of its parent universe, through to its birth from a black hole and so on stretching backward in time, potentially into infinity. Not only that, but, since our universe is filled with an estimated 100 million black holes, he proposes that each is the progenitor of one of 100 million universes that have descended from our own.

With black holes acting as the reproductive cells, or gametes, of universes, Smolin's model includes a kind of self-replication process. The next ingredient for his scenario is heredity. Smolin includes this by proposing that each offspring universe inherits the parameters, the values of the fundamental constants, particle masses and so on, of its parent. We could imagine these as cosmological genes* that encode the properties of universes just as biological genes encode the properties of living creatures.

Lastly, Smolin had to solve the problem that bedevilled Darwin and Wallace's theory of natural selection: finding a source of novel variation on which natural selection can cut its teeth. Inspired once again by biology, Smolin proposed that the tumultuous passage of a universe and its cosmic *genes* through a black hole sometimes causes their precise values to be altered by something analogous to mutation.

The idea that the laws of physics might be subject to change is not new. Smolin points out that the nineteenth-century American philosopher Charles Sanders Peirce (1839–1914), who was profoundly influenced by Darwinism, proposed that the laws of physics might, like living organisms, evolve. The English mathematician and philosopher William Kingdon Clifford (1845–79) made a similar claim. Even medieval theologians, such as William of Occam, argued that God might have created worlds different from our own. The physicists John Archibald Wheeler, Richard Feynman and Seth Lloyd have all proposed that the laws of physics

* I do realise that equating the parameters of the Standard Model with genes is stretching the analogy perhaps further than it can tolerate. But it does also pose some interesting questions. For example, if genetic information in genes is written in DNA, then what is the substrate for the universe's genes?

may be subject to change[10] both in space and time. Yet the case for changing laws within our own universe is weak because, as far as we can tell, precisely the same laws operate, both at its earliest moments and the furthest edges of our universe. Nevertheless, as Smolin points out, that does not stop the laws of physics being altered in different universes.

The starting point for the Smolin theory is a kind of cosmological equivalent of biology's origin of life scenario in which, at some time in the remote past, there was absolutely nothing. However, one of the most bizarre features of quantum mechanics is that we can't even be sure of nothing. This is another peculiar consequence of Heisenberg's uncertainty principle, which prevents us from being certain that even an absolute vacuum is empty of mass or energy. Quantum mechanics thereby leaves existential space for virtual particles to pop in and out of existence, even in the vacuum of empty space. In 1982, the Russian-born cosmologist Alexander Vilenkin made the even more astonishing claim that the universe arose in the simplest possible manner, as a quantum fluctuation 'from nothing'.[11]

Currently, the most accepted scenario for the universe's origin starts with a random quantum fluctuation. Most likely this tiny universe was not interesting as the randomly selected values of its fundamental constants were incompatible with even the existence of matter. As quickly as it popped into existence, the positive and negative energy of this matter-infeasible universe would have recombined to vanish into nothingness. Yet further quantum fluctuations would have continued to generate universes out of the cosmological nothingness until, after maybe many trillions of searches through parameter space, a universe was born with values that promoted the formation of matter, stars, planets and, at least a few, black holes.

The formation of black holes is, in Smolin's scenario, the cosmological equivalent of the origin of life: it made universes self-replicative. The early primordial universes would probably have generated few black holes and thereby produced only a handful of descendants. However, so long as that handful was greater than one, then the number of universes would increase. Moreover, because passage of the fundamental constants through the black hole is proposed to mutate their values, the descendant multiverse would have diversified into a kind of ecosystem of universes with different kinds of matter, stars, planets and variable numbers of black holes.

However, all universes would not be created equal. Some would be more fecund than others. Those that inherited parameter values that led to the biggest concentration of matter inside stars that collapsed into black holes would leave many descendants. Conversely, any universe that failed to produce stars or black holes would eventually blink out of existence to become extinct. Gradually, over a great many cosmological generations, the multiverse would be dominated by the fittest and most fecund universes underpinned by the rare values of the fundamental constants that generate the maximum number of black holes. Just as natural selection guided the evolution of life towards improbable creatures like dinosaurs, elephants or humans, so that same process, cosmological natural selection, fine-tuned the fundamental constants of physics towards the highly improbable values needed to make stars, planets, black holes and us.

This is a very remarkable theory but it is also, of course, very hard to prove. Nevertheless, the theory makes several predictions that Smolin has investigated in computer simulations of universe evolution. For example, the theory predicts that our universe, being the descendant of previously successful universes, should be fine-tuned to make abundant black holes such that any change in those values should result in fewer black holes. He tested this prediction by constructing a computer model of our universe and its fundamental constants and then tweaked their values. He discovered that even small changes to the parameters of the Standard Model either reduced the predicted number of black holes or resulted in no change. None of the computer model cosmological mutations were predicted to result in more black holes. His analysis suggests that our universe is indeed close to being as good as it gets at making black holes, just as cosmological natural selection predicts.

## The cosmological razor?

Although Smolin's cosmological natural selection theory might account for the fine-tuning of the fundamental constants, it does not, on its own, account for why, as Neil Turok argues, 'the universe has turned out to be stunningly simple'. When allied to Occam's razor, it just might.

First, I should emphasise that the following argument is not part of Smolin's CNS theory. Indeed, Smolin himself is sceptical that the world is close to being as simple it could be. He points at features that we have already discussed; for example, that two of the three generations of matter particles appear to be superfluous. I have already provided some possible roles for these extra particles. Another may be that they are required on symmetry grounds, being the cosmological equivalent of male nipples: not needed but hard to eliminate. Also, since generation II and III particles are rare, usually only found in particle accelerators or cosmic rays, none are likely to contribute to, or subtract from, black hole formation. They may be invisible to cosmological natural selection in the same way that the pseudogenes of naked mole-rats have dropped out of visibility to biological natural selection.

To see how a cosmological razor might evolve maximally simple universes, let's imagine that our universe contains two black holes that are the proud parents of two baby universes. When the fundamental constants pass through the first black hole, they emerge unscathed, encoding precisely the same values. Its descendants will thereby inherit identical constants to those in our own universe and will construct atoms and stars out of the seventeen particles of the Standard Model to produce a universe very like our own, which we will call the 17P universe reflecting its seventeen fundamental particles.

Imagine that when those same fundamental constants pass through the other black hole they suffer a cosmological mutation that spawns the genesis of, not only the seventeen particles of the Standard Model, but also an additional vestigial (roughly analogous to the vestigial limbs of a whale or the human appendix) eighteenth particle in this 18P baby universe. This extra particle does not do anything useful but instead merely hangs around, perhaps in inter-galactic clouds. The eighteenth particle is thereby an entity beyond necessity in the 18P universe.

While it plays no role in the formation of stars, black holes or people, the eighteenth particle will nevertheless influence their formation by robbing them of some of their mass. Let's hypothesise that it has average mass and abundance for fundamental particles so that it accounts for about one eighteenth of the total mass of the 18P universe. This locking away of mass in the intergalactic clouds of particle 18 will diminish the amount of matter/energy available for black-hole formation and thereby reduce the number of black holes by one eighteenth, or about 5 per cent. Since black holes are the mothers of universes, the 18P universe will generate around 5 per cent fewer progeny than its sibling 17P universe. If nothing else changes, then this difference in fecundity will continue into succeeding generations until, by about generation twenty, descendants of the 18P universe will be one third as abundant as descendants of the more parsimonious 17P universe. In the natural world, mutations that lead to just a 1 per cent reduction in fitness are enough to drive a mutant into extinction, so a 5 per cent decrease in fitness is likely to eliminate, or at least drastically reduce, the abundance of 18 particle universes relative to the more parsimonious 17 particle universes.

It is not clear whether the kind of multiverse envisaged by Smolin's theory is finite or infinite. If infinite, then the simplest universe capable of forming black holes will be infinitely more abundant than the next simplest universe. In this infinity of universes, we can ask the question, which one are we likely to inhabit? The answer is that we are infinitely more likely to inhabit the simplest possible universe that is compatible with life.

If instead the supply of universes is finite, then we have a similar situation to biological evolution on earth. Universes will compete for available resources – matter and energy – and the simplest that convert more of their mass into black holes will leave the most descendants. Once again, if we ask which universe we are most likely to inhabit, it will be the simplest. When inhabitants of these universes, like Robert Wilson and Arno Penzias, peer into the heavens to discover their cosmic microwave background and perceive its incredible smoothness, they, like Neil Turok, will remain baffled at how their universe has managed to do so much from such a 'stunningly simple' beginning. In short, they will inhabit a universe much like our own.

Smolin's theory, with or without its Occamist tweak, has one final and startling implication. It suggests that the fundamental law of the universe is not quantum mechanics, or general relativity or even the laws of mathematics. It is the law of natural selection discovered by Charles Darwin and Alfred Russel Wallace. As the philosopher Daniel Dennett insisted, it is 'The single best idea that anyone has ever had.'[12] It may also be the simplest idea that any universe has ever had.

# Epilogue

From the time he arrived in Munich around 1329, William of Occam remained under the protection of the Holy Roman Emperor Louis in or around Munich until his death aged about sixty. During that time the separatist Franciscans continued to hold the order's official seal of office and maintained a kind of government in exile. They also continued to write political treatises condemning Pope John XXII and successive popes, as well as commenting on a wide range of political issues, particularly the limits of papal and monarchic power. Sympathetic scholars visited the exiles, including scribes eager to copy their writings. Conrad de Vipeth's copy of Occam's *Summa Logica* with its portrait of William dates from this period. It now sits in the library of Gonville and Caius College, Cambridge. In a catalogue on the influence of the Franciscans in English art, it is annotated as having 'no value'.

As a contemporary political force, the rebel movement instigated by William and his fellow Franciscans failed. Pope John XXII provided a new official seal of office to the Franciscans and appointed a new and compliant minister general who went on to extinguish dissent in the order. Nevertheless, as we have discovered, despite the condemnations and bans, William's ideas rumbled through the medieval world, surviving the Black Death to resurface, usually without attribution, in the Renaissance, the Reformation and the Enlightenment.

Sadly, we know little of the later years of William's life, although we do know he sometimes travelled to European cities to lecture, still pursued by the agents of several popes. One pope even threatened to burn down the town of Tournai, in present-day Belgium, simply to capture the fugitive scholar.[1]

In 1342, when Michael of Cesena died, William became the keeper of the Franciscans' seal, guarding it until his death on 10 April 1347, the year the plague raged through Europe. William was buried, probably with the Franciscan seal, in the Church of St Francis in Munich. An inscription marking his grave stood on the site until the church was pulled down in 1803 and the tombstone destroyed. An opera house now stands there. The busy street of Occamstrasse lies nearby, hosting a Hotel Occam and a very convivial deli and wine bar called Occam Deli.

The deli is on the corner of Occamstrasse and Feilitzschstrasse. Walk a few minutes along Feilitzschstrasse and turn left into Leopoldstrasse and, after ten minutes or so, you will find yourself beneath the doors of the Ludwig Maximilian University of Munich where, between 1874 and 1877, Max Planck studied physics. You will remember that quantum mechanics dates its birth to 19 October 1900 when Planck presented his equation that blew away one of Lord Kelvin's clouds. Like all major scientific advances, it involved a simplification. At low frequencies, the Rayleigh–Jeans law correctly predicted the black body radiation whereas Wien's law was needed to predict the spectrum at high frequencies. Planck disproved neither law but showed that his revolutionary equation worked for the full range of frequencies. As the Nobel Prize-winning chemist Roald Hoffmann put it, Planck followed 'the logic, an Occam's Razor logic, to the quantum hypothesis'.[2]

Of course, Max Planck cited neither William of Occam nor his razor in his revolutionary paper. He did not need to. By this time, all scientists took their preference for simple solutions for granted despite the fact that most would have been hard-pressed to justify it. Opting for a complex theory when a simpler one will do the job is, to any modern scientist, simply unscientific. Yet, as we have discovered, this steadfast preference for simplicity within science is a relatively recent innovation and owes everything to William of Occam who blew away the dusty cobwebs of medieval doctrine to provide space for a leaner and sharper science. With Occam's razor in hand, that science has proved its worth by making sense of a baffling universe and providing us with happier, longer, healthier and more fulfilling lives than those enjoyed, or endured, by the majority of our ancestors.

I often think of William of Occam when I go for an early morning run on Wimbledon Common in south-west London, close to where I live. The path I usually take leads alongside Beverley Brook through Fishponds Wood, passing several small ponds that were once the fishponds of Merton Priory that was established in 1117 by Augustinian monks. It was still thriving in the early fourteenth century when William of Occam was living and studying around a three-hour walk or an hour's ride away at Greyfriars in London. I wonder if in the five or so years that William spent at Greyfriars he occasionally travelled south to visit the famous priory, perhaps even took part in a stormy disputation over scientific theology or the stark implications of God's omnipotence. If the weather was fine, he might have lingered a while to enjoy a wander by the fields, brook, ponds and woods as a refreshing break from the busy narrow streets and noxious odours of the Shambles surrounding his own friary.

Perhaps, like me on my morning run, he might have occasionally taken a wrong turn in the woods and lost the path. Eight centuries later, I found myself in precisely this situation (photograph on the left). With no clue to the direction needed to escape the thick undergrowth, progress would be impossible. But if, like me, he took a couple of steps forward and turned to the right, he would have discovered that he was already on a clear path out of the woods (photograph on the right). Just a change of perspective can be sufficient to allow us to escape the thickets of complexity and emerge into a simpler and more reasonable world.

Occam's razor is everywhere. It has cut a path through the thickets of misconceptions, dogmas, bigotries, biases, creeds, false beliefs and sheer baloney that has hindered progress in most times and most places. It is not that simplicity has been incorporated into modern science; simplicity is modern science and, through it, the modern world. Further simplicities are surely waiting to be discovered by scientists, particularly those from a much wider base of gender, race or sexual orientation who are unencumbered by the obvious prejudice, dogma and disadvantage that has restricted the story so far. And there is lots more work to be done even in the heartland of Occam's razor, physics, where no one has yet found a way of unifying the best theory we have for the structure of the cosmos, general relativity, with the best theory we have for the structure of atoms, quantum mechanics. As one of the greatest physicists of the twentieth century John Archibald Wheeler exclaimed: 'Behind it all is surely an idea so simple, so beautiful, that when we grasp it – in a decade, a century, or a millennium – we will all say to each other, how could it have been otherwise?'[3]

Life is indeed simple.

# Acknowledgements

I would like to thank all of those who very kindly read, and commented on, some or all the drafts of *Life is Simple* including (in no particular order) Sharon Kaye, Michael Brooks, John Gribbin, Bernard V. Lightman, Jim Al-Khalili, Jenny Pelletier, Rondo Keele, Mark Pallen, Philip Pullman, Michelle Collins, Tom McLeish, Patricia Fara, Jennifer Deane, Seb Falk, Robin Headlam Wells, Sara L. Uckelman, Greg Knowles, Tanya Baron, Philip Kim, Axel Theorell.

I would particularly like to thank my agent Patrick Walsh together with his wonderful team for keeping faith in the story of William of Occam and his razor through very trying times. Lastly, I would like to thank my brilliant editors Jamie Colman, Sarah Caro, Caroline Westmore and Martin Bryant, whilst reserving for myself the responsibility for any remaining errors.

# Credits

Figure 1: Gado Images/Alamy Stock Photo. Figure 2: NASA/WMAP Science Team WMAP # 121238. Figure 4: Mars motion 2018.png/ Tomruen/Creative Commons Attribution-ShareAlike 4.0 International (CC BY-SA 4.0). Figures 6, 11 and 21: Granger Historical Picture Archive/ Alamy Stock Photo. Figure 10: Classic Image/Alamy Stock Photo. Figures 13 below and 26: Interfoto/Alamy Stock Image. Figure 22: Louis Fibuler, *Ocean World: Being a Descriptive History of the Sea and its Living Inhabitants*, 1868, Plate XXVII. Image courtesy of Freshwater and Marine Image Bank/University of Washington Libraries. Figure 23: Paulo Oliveira/Alamy Stock Photo. Figure 24: Jacob van Maerlant, *Der Naturen Bloeme*, c.1350, photo Darling Archive/Alamy Stock Photo. Figure 27: The Natural History Museum/Alamy Stock Photo. Figure 28: Various ammonite fossils illustrating Hooke's discourse on Earthquakes. Wellcome Collection/ Q5QW2R83. Creative Commons Attribution licence 4.0 International (CC BY 4.0). Figure 29: Henry Walter Bates, *The Naturalist on the River Amazon*, vol. 1, 1863, Frontispiece. Figure 30: Andrew Wood/Alamy Stock Photos. Figure 32: Naked mole rat-National Museum of Nature and Science.jpg Momotarou2012/Creative Commons Attribution-ShareAlike 3.0 Unported (CC BY-SA 3.0). Figure 34: courtesy Dr G. Breitenbach. Figure 35: Ian Dagnall Computing/Alamy Stock Photo. Figures 36 and 42: Author's collection. Figure 38: Shutterstock.com (six-sided dice); D60 60men-saikoro.jpg/Saharasav/Creative Commons Attribution-ShareAlike 4.0 International (CC BY-SA 4.0) (sixty-sided dice).

Figures 3, 5, 7, 8, 9, 12, 13 above, 14, 15, 16, 17, 18, 19, 20, 25, 31, 33, 37, 39, 40 and 41 drawn by Penny Amerena.

# Notes

## Introduction

1. Wilkinson, D. T., and Peebles, P., in *Particle Physics and the Universe*, 136–41 (World Scientific, 2001).
2. Turok, N., 'The Astonishing Simplicity of Everything', public lecture at the Perimeter Institute for Theoretical Physics, Ontario, Canada, 7 October 2015, https://www.youtube.com/watch?v=f1x-9lgX8GaE.
3. Sober, E., *Ockham's Razors* (Cambridge University Press, 2015).
4. Doyle, A. C., *The Sign of Four* (Broadview Press, 2010).
5. Barnett, L., and Einstein, A., *The Universe and Dr Einstein* (Courier Corporation, 2005).
6. Wootton, D., *The Invention of Science: A New History of the Scientific Revolution* (Penguin, 2015). Gribbin, J., *Science: A History* (Penguin, 2003). Ignotofsky, R., *Women in Science: 50 Fearless Pioneers Who Changed the World* (Ten Speed Press, 2016). Kuhn, T. S., *The Structure of Scientific Revolutions* (University of Chicago Press, 2012).

## Chapter 1: Of Scholars and Heretics

1. de Ockham, G., and Ockham, W., *William of Ockham: 'A Letter to the Friars Minor' and Other Writings* (Cambridge University Press, 1995).
2. Knysh, G., 'Biographical Rectifications Concerning Ockham's Avignon Period', *Franciscan Studies*, 46, 61–91 (1986).
3. Villehardouin, G., and De Joinville, J., *Chronicles of the Crusades* (Courier Corporation, 2012).

4. Evans, J., *Life in Medieval France* (Phaidon Paperback, 1957).

5. Sparavigna, A. C., 'The Light Linking Dante Alighieri to Robert Grosseteste', PHILICA, Article no. 572 (2016).

6. Gill, M. J., *Angels and the Order of Heaven in Medieval and Renaissance Italy* (Cambridge University Press, 2014).

7. Jowett, B., and Campbell, L., *Plato's Republic*, vol. 3, 518c (Clarendon Press, 1894).

8. Smith, A. M., 'Saving the Appearances of the Appearances: The Foundations of Classical Geometrical Optics', *Archive for History of Exact Sciences*, 73–99 (1981).

9. Deakin, M. A., 'Hypatia and Her Mathematics', *American Mathematical Monthly*, 101, 234–43 (1994).

10. Charles, R. H., *The Chronicle of John, Bishop of Nikiu: Translated from Zotenberg's Ethiopic Text*, vol. 4 (Arx Publishing, 2007).

11. Munitz, M. K., *Theories of the Universe* (Simon and Schuster, 2008).

## Chapter 2: The Physics of God

1. Laistner, M., 'The Revival of Greek in Western Europe in the Carolingian Age', *History*, 9, 177–87 (1924).

2. Clark, G., 'Growth or Stagnation? Farming in England, 1200–1800', *Economic History Review*, 71, 55–81 (2018).

3. Nordlund, T., 'The Physics of Augustine: The Matter of Time, Change and an Unchanging God', *Religions*, 6, 221–44 (2015).

4. Gill, T., *Confessions* (Bridge Logos Foundation, 2003).

5. Al-Khalili, J., *Pathfinders: The Golden Age of Arabic Science* (Penguin, 2010).

6. Dinkova-Bruun, G., et al., *The Dimensions of Colour: Robert Grosseteste's De Colore* (Institute of Medieval and Renaissance Studies, 2013).

7. Hannam, J., *The Genesis of Science: How the Christian Middle Ages Launched the Scientific Revolution* (Regnery Publishing, 2011).

8. Zajonc, A., *Catching the Light: The Entwined History of Light and Mind* (Oxford University Press, USA, 1995).

9. Meri, J. W., *Medieval Islamic Civilization: An Encyclopedia* (Routledge, 2005).

10. Lombard, P., *The First Book of Sentences on the Trinity and Unity of God*; https://franciscan-archive.org/lombardus/I-Sent.html.

11. Sylla, E. D., in *The Cultural Context of Medieval Learning*, 349–96 (Springer, 1975).
12. Riddell, J., *The Apology of Plato* (Clarendon Press, 1867).

## Chapter 3: The Razor

1. Hammer, C. I., 'Patterns of Homicide in a Medieval University Town: Fourteenth-Century Oxford', *Past & Present*, 3–23 (1978).
2. Ibid.
3. Little, A. G., *Franciscan History and Legend in English Mediaeval Art*, vol. 19 (Manchester University Press, 1937).
4. Lambertini, R., 'Francis of Marchia and William of Ockham: Fragments From a Dialogue', *Vivarium*, 44, 184–204 (2006).
5. Leff, G., *William of Ockham: The Metamorphosis of Scholastic Discourse* (Manchester University Press, 1975).
6. Tornay, S. C., 'William of Ockham's Nominalism', *Philosophical Review*, 45, 245–67 (1936).
7. Ibid.
8. Loux, M. J., *Ockham's Theory of Terms: Part I of the Summa Logicae* (St Augustine's Press, 2011).
9. Goddu, A., *The Physics of William of Ockham*, vol. 16 (Brill Archive, 1984).
10. Freddoso, A. J., *Quodlibetal Questions* (Yale University Press, 1991).
11. Kaye, S. M., and Martin, R. M., *On Ockham* (Wadsworth/Thompson Learning Inc., 2001).
12. Sylla, E. D., in *The Cultural Context of Medieval Learning*, 349–96 (Springer, 1975).
13. Shea, W. R., 'Causality and Scientific Explanation, Vol. I: Medieval and Early Classical Science by William A. Wallace', *Thomist: A Speculative Quarterly Review*, 37, 393–6 (1973).
14. Leff, *William of Ockham*.
15. Spade, P. V., *The Cambridge Companion to Ockham* (Cambridge University Press, 1999).
16. Ibid.
17. Kaye and Martin, *On Ockham*.
18. Keele, R., *Ockham Explained: From Razor to Rebellion*, vol. 7 (Open Court Publishing, 2010).
19. de Ockham, G., and Ockham, W., *William of Ockham: 'A Letter to the*

*Friars Minor' and Other Writings* (Cambridge University Press, 1995).

20. Mollat, G., *The Popes at Avignon: 1305–1378* (trans. J. Love), 38–9 (Thomas Nelson & Sons, 1963).

21. Brampton, C. K., 'Personalities at the Process Against Ockham at Avignon, 1324–26', *Franciscan Studies*, 26, 4–25 (1966). Birch, T. B., *The De Sacramento Altaris of William of Ockham* (Wipf and Stock Publishers, 2009).

## Chapter 4: How Simple Are Rights?

1. Van Duffel, S., and Robertson, S., *Ockham's Theory of Natural Rights* (available at SSRN 1632452, 2010).

2. Deane, J. K., *A History of Medieval Heresy and Inquisition* (Rowman & Littlefield Publishers, 2011).

3. Mariotti, L., *A Historical Memoir of Frà Dolcino and his Times; Being an Account of a general Struggle for Ecclesiastical Reform and of an anti-heretical crusade in Italy, in the early part of the fourteenth century* (Longman, Brown, Green, 1853).

4. Burr, D., *The Spiritual Franciscans: From Protest to Persecution in the Century After Saint Francis* (Penn State Press, 2001).

5. Haft, A. J., White, J. G., and White, R. J., *The Key to 'The Name of the Rose': Including Translations of All Non-English Passages* (University of Michigan Press, 1999).

6. de Ockham, G., and Ockham, W., *William of Ockham: 'A Letter to the Friars Minor' and Other Writings* (Cambridge University Press, 1995).

7. Knysh, G., 'Biographical Rectifications Concerning Ockham's Avignon Period', *Franciscan Studies*, 46, 61–91 (1986).

8. Leff, G., *William of Ockham: The Metamorphosis of Scholastic Discourse* (Manchester University Press, 1975).

9. Tierney, B., *The Idea of Natural Rights: Studies on Natural Rights, Natural Law, and Church Law, 1150–1625*, vol. 5 (Wm. B. Eerdmans Publishing, 2001).

10. Tierney, B., 'The Idea of Natural Rights-Origins and Persistence', *Northwestern Journal of International Human Rights*, 2, 2 (2004).

11. Ibid.

12. Witte Jr, J., and Van der Vyver, J. D., *Religious Human Rights in*

*Global Perspective: Religious Perspectives*, vol. 2 (Wm. B. Eerdmans Publishing, 1996).

13. Tierney, B., 'Villey, Ockham and the Origin of Individual Rights', in Witte, J. (ed.), *The Weightier Matters of the Law: Essays on Law and Religion*, 1–31 (Scholars Press, 1988).

14. Chroust, A.-H., 'Hugo Grotius and the Scholastic Natural Law Tradition', *New Scholasticism*, 17, 101–33 (1943).

15. Trachtenberg, O., 'William of Occam and the Prehistory of English Materialism', *Philosophy and Phenomenological Research*, 6, 212–24 (1945).

## Chapter 5: The Kindling

1. Etzkorn, G. J., 'Codex Merton 284: Evidence of Ockham's Early Influence in Oxford', *Studies in Church History Subsidia*, 5, 31–42 (1987).

2. Aleksander, J., 'The Significance of the Erosion of the Prohibition against Metabasis to the Success and Legacy of the Copernican Revolution', *Annales Philosophici*, 9–22 (2011).

3. McGinnis, J., 'A Medieval Arabic Analysis of Motion at an Instant: The Avicennan Sources to the forma fluens/fluxus formae Debate', *British Journal for the History of Science*, 39(2), 189–205 (2006).

4. Copleston, F., *A History of Philosophy*, vol. 3: *Ockham to Suarez* (Paulist Press, 1954).

5. Goddu, A., 'The Impact of Ockham's Reading of the Physics on the Mertonians and Parisian Terminists', *Early Science and Medicine*, 6, 204–36 (2001).

6. Sylla, E. D., 'Medieval Dynamics', *Physics Today*, 61, 51 (2008).

7. Courtenay, W. J., 'The Reception of Ockham's Thought in Fourteenth-Century England', *From Ockham to Wyclif*, 89–107 (Boydell and Brewer, 1987).

8. Goddu, 'The Impact of Ockham's Reading of the Physics'.

9. Heytesbury, W., *On Maxima and Minima: Chapter 5 of Rules for Solving Sophismata: With an Anonymous Fourteenth-Century Discussion*, vol. 26 (Springer Science & Business Media, 2012).

10. Wikipedia definition of speed, https://en.wikipedia.org/wiki/Speed#Historical_definition.

11. Barnett, L., and Einstein, A., *The Universe and Dr Einstein* (Courier Corporation, 2005).

12. Klima, G., *John Buridan* (Oxford University Press, 2008).
13. Goddu, A., *The Physics of William of Ockham*, vol. 16 (Brill Archive, 1984).
14. Tachau, K., *Vision and Certitude in the Age of Ockham: Optics, Epistemology and the Foundation of Semantics 1250–1345* (Brill, 2000).
15. Hannam, J., *God's Philosophers: How the Medieval World Laid the Foundations of Modern Science* (Icon Books, 2009).
16. Ibid.
17. Shapiro, H., *Medieval Philosophy: Selected Readings from Augustine to Buridan* (Modern Library, 1964).

## Chapter 6: The Interregnum

1. Alfani, G., and Murphy, T. E., 'Plague and Lethal Epidemics in the Pre-Industrial World', *Journal of Economic History*, 77, 314–43 (2017).
2. Nicholl, C., *Leonardo da Vinci: The Flights of the Mind* (Penguin, 2005).
3. Ibid.
4. Ibid.
5. Reti, L., 'The Two Unpublished Manuscripts of Leonardo da Vinci in the Biblioteca Nacional of Madrid – II', *Burlington Magazine*, 110, 81–91 (1968).
6. Duhem, P., 'Research on the History of Physical Theories', *Synthese*, 83, 189–200 (1990).
7. Randall, J. H., 'The Place of Leonardo Da Vinci in the Emergence of Modern Science', *Journal of the History of Ideas*, 191–202 (1953).
8. Long, M. P., 'Francesco Landini and the Florentine Cultural Elite', *Early Music History*, 3, 83–99 (1983).
9. Funkenstein, A., *Theology and the Scientific Imagination: From the Middle Ages to the Seventeenth Century* (Princeton University Press, 2018).
10. Matsen, H., 'Alessandro Achillini (1463–1512) and "Ockhamism" at Bologna (1490–1500)', *Journal of the History of Philosophy*, 13, 437–51 (1975).
11. Dutton, B. D., 'Nicholas of Autrecourt and William of Ockham on Atomism, Nominalism, and the Ontology of Motion', *Medieval Philosophy & Theology*, 5, 63–85 (1996).

12. Reti, 'The Two Unpublished Manuscripts of Leonardo da Vinci'.

13. Gillespie, M. A., *Nihilism Before Nietzsche* (University of Chicago Press, 1995).

14. Ibid.

15. Boysen, B., 'The Triumph of Exile: The Ruptures and Transformations of Exile in Petrarch', *Comparative Literature Studies*, 55, 483–511 (2018).

16. Petrarca, F., 'On His Own Ignorance and That of Many Others' (trans. Hans Nicod), in Cassirer, E., Kristeller, P. O., and Randall, J. H. (eds), in *The Renaissance Philosophy of Man: Petrarca, Valla, Ficino, Pico, Pomponazzi, Vives*, 47–133 (University of Chicago Press, 2011).

17. Medieval Sourcebook: Petrarch, *The Ascent of Mount Ventoux*; https://sourcebooks.fordham.edu/source/petrarch-ventoux.asp.

18. Rawski, C. H., *Petrarch's Remedies for Fortune Fair and Foul: A Modern English Translation of De Remediis Utriusque Fortune, With A Commentary. References: Bibliography, Indexes, Tables and Maps*, vol. 2, 226 (Indiana University Press, 1991).

19. Trinkaus, C., 'Petrarch's Views on the Individual and His Society', *Osiris*, 11, 168–98 (1954).

20. Boucher, H. W., 'Nominalism: The Difference for Chaucer and Boccaccio', *Chaucer Review*, 213–20 (1986).

21. Keiper, H., Bode, C., and Utz, R. J., *Nominalism and Literary Discourse: New Perspectives*, vol. 10 (Rodopi, 1997).

22. Dvořák, M. *The History of Art as a History of Ideas* (trans. John Hardy) (Routledge & Kegan Paul, 1984).

23. Hauser, A., *The Social History of Art*, vol. 2: *Renaissance* (Routledge, 2005).

24. Kieckhefer, R., *Magic in the Middle Ages* (Cambridge University Press, 2000).

25. Holborn, H., *A History of Modern Germany: The Reformation*, vol. 1 (Princeton University Press, 1982).

26. Oberman, H., *The Dawn of the Reformation: Essays in Late Medieval and Early Reformation Thought* (Wm. B. Eerdmans Publishing, 1992).

27. Gillespie, M. A., *The Theological Origins of Modernity* (University of Chicago Press, 2008).

28. Pekka, K., in *Encyclopedia of Medieval Philosophy: Philosophy Between 500 and 1500* (ed. Henrik Lagerlund), 14–45 (Springer, 2011).

## Chapter 7: The Heliocentric but Hermetic Cosmos

1. Krauze-Błachowicz, K., 'Was Conceptualist Grammar in Use at Cracow University?', *Studia Antyczne i Mediewistyczne*, 6, 275–85 (2008).
2. Matsen, H., 'Alessandro Achillini (1463–1512) and "Ockhamism" at Bologna (1490–1500)', *Journal of the History of Philosophy*, 13, 437–51 (1975).
3. Edelheit, A., *Ficino, Pico and Savonarola: The Evolution of Humanist Theology 1461/2–1498* (Brill, 2008).
4. Barbour, J. B., *The Discovery of Dynamics: A Study from a Machian Point of View of the Discovery and the Structure of Dynamical Theories* (Oxford University Press, 2001).
5. Sobel, D., *A More Perfect Heaven: How Copernicus Revolutionised the Cosmos*, 178 (A & C Black, 2011).
6. Ibid.
7. Gingerich, O., '"Crisis" Versus Aesthetic in the Copernican Revolution', *Vistas in Astronomy*, 17, 85–95 (1975).
8. Copernicus, N., *On the Revolutions* (trans. and commentary Edward Rosen) (Johns Hopkins University Press, 1978), http://www.geo.utexas.edu/courses/302d/Fall_2011/Full%20text%20-%20 Nicholas%20Copernicus,%20_De%20Revolutionibus%20(On%20 the%20Revolutions),_%201.pdf.
9. Gingerich, O., '"Crisis" Versus Aesthetic in the Copernican Revolution', *Vistas in Astronomy*, 17, 85–95 (1975).

## Chapter 8: Breaking the Spheres

1. Thoren, V. E., *The Lord of Uraniborg: A Biography of Tycho Brahe* (Cambridge University Press, 1990).
2. Ibid.
3. Oberman, H. A., *The Harvest of Medieval Theology: Gabriel Biel and Late Medieval Nominalism* (Harvard University Press, 1963).
4. Methuen, C., *Kepler's Tübingen: Stimulus to a Theological Mathematics* (Ashgate, 1998).
5. Field, J. V., 'A Lutheran Astrologer: Johannes Kepler', *Archive for History of Exact Sciences*, 31 (1984).
6. Spielvogel, J. J., *Western Civilization*, 467 (Cengage Learning, 2014).
7. Bialas, V., *Johannes Kepler*, vol. 566 (CH Beck, 2004).
8. Chandrasekhar, S., *The Pursuit of Science*, 410–20 (Minerva, 1984).

9. Kepler, J., *The Harmony of the World*, vol. 209, 302 (American Philosophical Society, 1997).

10. Poincaré, H., and Maitland, F., *Science and Method* (Courier Corporation, 2003).

11. Dirac, P. A. M., 'XI. – The Relation Between Mathematics and Physics', *Proceedings of the Royal Society of Edinburgh*, 59, 122–9 (1940).

12. Kepler, *The Harmony of the World*.

13. Martens, R., *Kepler's Philosophy and the New Astronomy* (Princeton University Press, 2000).

14. Sober, E., *Ockham's Razors* (Cambridge University Press, 2015). Sober, E., 'What is the Problem of Simplicity', *Simplicity, Inference, and Econometric Modelling*, 13–32 (2002).

15. Fraser, J., 'The Ever-Presence of Eternity', *Dialog*, 39, 40–5 (2000).

## Chapter 9: Bringing Simplicity Down to Earth

1. Sober, E., *Ockham's Razors* (Cambridge University Press, 2015).

2. Reeves, E. A., *Galileo's Glassworks: The Telescope and the Mirror* (Harvard University Press, 2009).

3. Ibid.

4. Galilei, G., and Van Helden, A., *Sidereus Nuncius, or the Sidereal Messenger* (University of Chicago Press, 2016).

5. Wootton, D., *Galileo: Watcher of the Skies* (Yale University Press, 2010).

6. Galilei, G., and Wallace, W. A., *Galileo's Early Notebooks: The Physical Questions: A Translation from the Latin, with Historical and Paleographical Commentary* (University of Notre Dame Press, 1977).

7. Buchwald, J. Z., *A Master of Science History: Essays in Honor of Charles Coulston Gillispie*, vol. 30 (Springer Science & Business Media, 2012).

8. Sober, *Ockham's Razors*.

## Chapter 10: Atoms and Knowing Spirits

1. Wojcik, J. W., *Robert Boyle and the Limits of Reason*, 151–88 (Cambridge University Press, 1997).

2. Hunter, M., *Boyle: Between God and Science* (Yale University Press, 2010).

3. Ibid.

4. Ibid.

5. Ibid.

6. Ibid.

7. Pilkington, R., *Robert Boyle: Father of Chemistry* (John Murray, 1959).

8. Descartes, R., *Discourse on the Method of Rightly Conducting the Reason, and Seeking Truth in the Sciences* (Sutherland and Knox, 1850).

9. Goddu, A., *The Physics of William of Ockham*, vol. 16 (Brill Archive, 1984).

10. Hull, G., 'Hobbes's Radical Nominalism', *Epoché: A Journal for the History of Philosophy*, 11, 201–23 (2006).

11. Hobbes, T., *Hobbes's Leviathan*, vol. 1 (Google Books, 1967).

12. Ibid.

13. Gillespie, M. A., *The Theological Origins of Modernity* (ReadHowYouWant.com, 2010).

14. Lindberg, D. C., and Numbers, R. L., *When Science and Christianity Meet* (University of Chicago Press, 2008).

15. Medawar, P., *The Art of the Soluble* (Methuen, 1967).

16. Milton, J. R., 'Induction Before Hume', *British Journal for the Philosophy of Science*, 38, 49–74 (1987).

17. Hunter, *Boyle*.

18. Greene, R. A., 'Henry More and Robert Boyle on the Spirit of Nature', *Journal of the History of Ideas*, 451–74 (1962)

19. Wojcik, *Robert Boyle*.

20. Ibid., 174.

21. Descartes, R., 'Rules for the Direction of the Mind', in *The Philosophical Works of Descartes* (trans. E. S. Haldane and G. R. T. Ross), vol. 1, 7 (Dover Publications, 1955).

22. Wood, A., and Bliss, P., *Athenæ Oxonienses: An Exact History of All the Writers and Bishops who Have Had Their Education in the University of Oxford. To which are Added, the Fasti Or Annals, of the Said University* (F. C. & J. Rivington, 1820).

## Chapter 11: The Notion of Motion

1. Stewart, L., 'Other Centres of Calculation, or, Where the Royal Society Didn't Count: Commerce, Coffee-Houses and Natural Philosophy in Early Modern London', *British Journal for the History of Science*, 32, 133–53 (1999).

2. Koyré, A., 'An Unpublished Letter of Robert Hooke to Isaac Newton', *Isis*, 43, 312–37 (1952).

3. Whitehead, A. N., *Principia mathematica* (1913).
4. Copleston, F., *A History of Philosophy*, vol. 3: *Ockham to Suarez* (Paulist Press, 1954).
5. Quoted in Kaye, S. M., and Martin, R. M., *On Ockham* (Wadsworth/ Thompson Learning, 2001).

## Chapter 12: Making Motion Work

1. Feuer, L. S., 'The Principle of Simplicity', *Philosophy of Science*, 24, 109–22 (1957).
2. Kitcher, P., *The Advancement of Science: Science Without Legend, Objectivity Without Illusions*, 280 (Oxford University Press on Demand, 1995).
3. Brown, S. C., 'Count Rumford and the Caloric Theory of Heat', *Proceedings of the American Philosophical Society*, 93, 316–25 (1949).

## Chapter 13: The Vital Spark

1. von Walde-Waldegg, H., 'Notes on the Indians of the Llanos of Casanare and San Martin (Colombia)', *Primitive Man*, 9, 38–45 (1936).
2. von Humboldt, A., Bonpland, A., and Ross, T., *Personal Narrative of Travels to the Equinoctial Regions of America: During the Years 1799– 1804*, vols 1–3 (G. Bell & Sons, 1894).
3. Lattman, P., 'The Origins of Justice Stewart's "I Know It When I See It"', *Wall Street Journal*, 27 September 2007, https://www.wsj. com/articles/BL-LB-4558.
4. Laertius, R. D. D., and Hicks, R. D., *Lives of Eminent Philosophers* (trans. R. D. Hicks) (Heinemann, 1959).
5. Finger, S., and Piccolino, M., *The Shocking History of Electric Fishes: From Ancient Epochs to the Birth of Modern Neurophysiology* (Oxford University Press USA, 2011).
6. Copenhaver, B. P., 'A Tale of Two Fishes: Magical Objects in Natural History from Antiquity Through the Scientific Revolution', *Journal of the History of Ideas*, 52, 373–98, doi:10.2307/2710043 (1991).
7. Ibid.
8. Finger and Piccolino, *The Shocking History of Electric Fishes*.
9. Solomon, S., et al., 'Safety and Effectiveness of Cranial

Electrotherapy in the Treatment of Tension Headache', *Headache*, 29, 445–50, doi:10.1111/j.1526-4610.1989.hed2907445.x (1989).

10. Copenhaver, 'A Tale of Two Fishes'.

11. Compagnon, A., *Nous: Michel de Montaigne* (Le Seuil, 2016).

12. Finger and Piccolino, *The Shocking History of Electric Fishes*.

13. Ibid.

14. Wulf, A., *The Invention of Nature: Alexander Von Humboldt's New World* (Alfred A. Knopf, 2015).

15. Finkelstein, G., *Emil Du Bois-Reymond: Neuroscience, Self, and Society in Nineteenth-Century Germany* (MIT Press, 2013).

16. Gorby, Y. A., et al., 'Electrically Conductive Bacterial Nanowires Produced by Shewanella oneidensis Strain MR-1 and Other Micro-organisms', *Proceedings of the National Academy of Sciences*, 103, 11358–63 (2006).

17. Vandenberg, L. N., Morrie, R. D., and Adams, D. S., 'V-ATPase-Dependent Ectodermal Voltage and pH Regionalization Are Required for Craniofacial Morphogenesis', *Developmental Dynamics*, 240, 1889–1904 (2011).

## Chapter 14: Life's Vital Direction

1. Wallace, A. R., *Darwinism: An Exposition of the Theory of Natural Selection with Some of its Applications* (Cosimo, Inc., 2007).

2. Kaye, S. M., *William of Ockham* (Oxford University Press, 2015).

3. Wallace Letters online, Natural History Museum, London; https://www.nhm.ac.uk/research-curation/scientific-resources/collections/library-collections/wallace-letters-online/index.html.

4. Wallace, A. R., 'On the Law Which Has Regulated the Introduction of New Species (1855)', *Alfred Russel Wallace Classic Writings*, Paper 2 (2009), http://digitalcommons.wku.edu/dlps_fac_arw/2

5. Ereshefsky, M., 'Some Problems with the Linnaean Hierarchy', *Philosophy of Science*, 61, 186–205 (1994).

6. Winchester, S., *The Map That Changed the World: A Tale of Rocks, Ruin and Redemption* (Penguin, 2002).

7. Ibid.

8. Goodhue, T. W., *Fossil Hunter: The Life and Times of Mary Anning (1799–1847)* (Academica Press, 2004).

9. Raby, P., *Alfred Russel Wallace: A Life* (Princeton University Press, 2002).

10. Bowler, P. J., *Evolution: The History of an Idea: 25th Anniversary Edition, With a New Preface* (University of California Press, 2009).
11. Raby, *Alfred Russel Wallace.*
12. Van Wyhe, J., 'The Impact of AR Wallace's Sarawak Law Paper Reassessed', *Studies in History and Philosophy of Science Part C: Studies in History and Philosophy of Biological and Biomedical Sciences*, 60, 56–66 (2016).
13. Ibid.
14. Davies, R., '1 July 1858: What Wallace Knew; What Lyell Thought He Knew; What Both He and Hooker Took on Trust; And What Charles Darwin Never Told Them', *Biological Journal of the Linnean Society*, 109, 725–36 (2013).
15. Beddall, B. G., 'Darwin and Divergence: The Wallace Connection', *Journal of the History of Biology*, 21.1, 1–68 (1988).
16. Shermer, M., *In Darwin's Shadow: The Life and Science of Alfred Russel Wallace: A Biographical Study on the Psychology of History* (Oxford University Press on Demand, 2002). Cowan, I., 'A Trumpery Affair: How Wallace Stimulated Darwin to Publish and Be Damned'; http://wallacefund.info/sites/wallacefund.info/files/A%20Trumpery%20Affair.pdf.
17. Kutschera, U., 'Wallace Pioneered Astrobiology Too', *Nature*, 489, 208 (2012).
18. Dennett, D. C., *Darwin's Dangerous Idea: Evolution and the Meanings of Life* (Simon & Schuster, 1996).
19. Wallace, A. R., and Berry, A., *The Malay Archipelago* (Penguin, 2014).

## Chapter 15: Of Peas, Primroses, Flies and Blind Rodents

1. Wallace, A. R., *Mimicry, and Other Protective Resemblances Among Animals* (Read Books Limited, 2016).
2. Vorzimmer, P., 'Charles Darwin and Blending Inheritance', *Isis*, 54, 371–90 (1963).
3. De Castro, M., 'Johann Gregor Mendel: Paragon of Experimental Science', *Molecular Genetics & Genomic Medicine*, 4, 3 (2016).
4. Mendel, G., *Experiments on Plant Hybrids* (1866), translation and commentary by Staffan Müller-Wille and Kersten Hall, British Society for the History of Science Translation Series (2016), http://www.bshs.org.uk/bshs-translations/mendel.

5. Ibid.

6. Dobzhansky, T., 'Nothing in Biology Makes Sense Except in the Light of Evolution', *American Biology Teacher*, 35, 125–9 (1973).

7. Bergson, H., *Creative Evolution*, vol. 231 (University Press of America, 1911).

8. Watson, J., *The Double Helix* (Weidenfeld & Nicolson, 2010). Maddox, B., *Rosalind Franklin: The Dark Lady of DNA* (HarperCollins New York, 2002). Watson, J. D., Berry, A., and Davies, K., *DNA: The Story of the Genetic Revolution* (Knopf, 2017).

9. Karafyllidis, I. G., 'Quantum Mechanical Model for Information Transfer From DNA to Protein', *Biosystems*, 93, 191–8 (2008).

10. Dawkins, R., *The Selfish Gene* (Oxford University Press, 1976).

11. Sherman, P. W., Jarvis, J. U., and Alexander, R. D., *The Biology of the Naked Mole-Rat* (Princeton University Press, 2017).

12. Kim, E. B., et al., 'Genome Sequencing Reveals Insights Into Physiology and Longevity of the Naked Mole Rat', *Nature*, 479, 223–7 (2011).

13. Meredith, R. W., Gatesy, J., Cheng, J., and Springer, M. S., 'Pseudogenization of the Tooth Gene Enamelysin (MMP20) in the Common Ancestor of Extant Baleen Whales', *Proceedings of the Royal Society of London B: Biological Sciences*, rspb20101280 (2010).

14. Zhao, H., Yang, J.-R., Xu, H., Zhang, J., 'Pseudogenization of the Umami Taste Receptor Gene Tas1r1 in the Giant Panda Coincided With Its Dietary Switch to Bamboo', *Molecular Biology and Evolution*, 27, 2669–73 (2010).

15. Li, X., et al., 'Pseudogenization of a Sweet-Receptor Gene Accounts for Cats' Indifference Toward Sugar', *PLoS Genetics*, 1 (2005).

## Chapter 16: The Best of All Possible Worlds?

1. Heisenberg, W., *Physics and Beyond: Encounters and Conversations* (1969) (HarperCollins, 1971).

2. Gribbin, J., *Science: A History* (Penguin, 2003). Gribbin, J., *Schrodinger's Kittens: And the Search for Reality* (Weidenfeld & Nicolson, 2012). Rovelli, C., *The Order of Time* (Riverhead, 2019). Fara, P., *Science: A Four Thousand Year History* (Oxford University Press, 2010). Cox, B., and Forshaw, J., *Why Does E= mc2?* (Da Capo, Boston, 2009). Al-Khalili, J., *The World According to Physics* (Princeton

University Press, 2020). Green, B., *The Fabric of the Cosmos: Space, Time, and the Texture of Reality* (Penguin, 2004).

3. Norton, J. D., 'Nature is the Realisation of the Simplest Conceivable Mathematical Ideas: Einstein and the Canon of Mathematical Simplicity', *Studies in History and Philosophy of Science Part B: Studies in History and Philosophy of Modern Physics*, 31, 135–70 (2000).

4. Ibid.

## Chapter 17: A Quantum of Simplicity

1. Betten, F. S., 'Review of: De Sacramento Altaris of William of Ockham by T. Bruce Birch', *Catholic Historical Review*, 20, 50–6 (1934).

2. McFadden, J., *Quantum Evolution* (HarperCollins, 2000).

3. Al-Khalili, J., and McFadden, J., *Life on the Edge: The Coming of Age of Quantum Biology* (Bantam Press, 2014).

4. Tent, M. B. W., *Emmy Noether: The Mother of Modern Algebra* (CRC Press, 2008).

5. Brewer, J. W., Noether, E., and Smith, M. K., *Emmy Noether: A Tribute to Her Life and Work* (Dekker, 1981).

6. Arntzenius, F., *Space, Time, and Stuff* (Oxford University Press, 2014). Chen, E. K., 'An Intrinsic Theory of Quantum Mechanics: Progress in Field's Nominalistic Program, Part I' (Oxford University Press, 2014).

7. Wigner, E. P., 'The Unreasonable Effectiveness of Mathematics in the Natural Sciences', *Communications on Pure and Applied Mathematics*, 13, 001–014 (1960).

## Chapter 18: Opening Up the Razor

1. Russell, B., *Our Knowledge of the External World* (Jovian Press, 2017).

2. Bellhouse, D. R., 'The Reverend Thomas Bayes, FRS: A Biography to Celebrate the Tercentenary of His Birth', *Statistical Science*, 19, 3–43 (2004).

3. Wilmott, J., *The Debt to Pleasure* (Carcanet, 2012).

4. Jeffreys, H., *The Theory of Probability* (Oxford University Press, 1998).

5. Gull, S. F., in *Maximum-Entropy and Bayesian Methods in Science and*

*Engineering*, 53–74 (Springer, 1988). Jefferys, W. H., and Berger, J. O., 'Sharpening Ockham's Razor on a Bayesian Strop', Technical Report (1991).

6. Sober, E., *Ockham's Razors* (Cambridge University Press, 2015).
7. Kuhn, T. S., *The Structure of Scientific Revolutions* (University of Chicago Press, 2012).
8. Koestler, A., *The Sleepwalkers: A History of Man's Changing Vision of the Universe* (Penguin, 2017).
9. Feyerabend, P., *Against Method* (Verso, 1993).
10. Kaye, S. M., and Martin, R. M., *On Ockham* (2001).
11. Kuhn, *The Structure of Scientific Revolutions*.
12. Rorty, R., *Contingency, Irony, and Solidarity* (Cambridge University Press, 1989).
13. Blaedel, N., *Harmony and Unity: The Life of Niels Bohr* (Science Tech. Publ., 1988).
14. Carey, N., *The Epigenetics Revolution: How Modern Biology is Rewriting Our Understanding of Genetics, Disease, and Inheritance* (Columbia University Press, 2012).
15. Chater, N., and Vitányi, P., 'Simplicity: A Unifying Principle in Cognitive Science?', *Trends in Cognitive Sciences*, 7, 19–22 (2003).
16. Goodman, N., 'The Test of Simplicity', *Science*, 128, 1064–9 (1958).

## Chapter 19: The Simplest of All Possible Worlds?

1. McCurdy, E., *The Notebooks of Leonardo da Vinci*, vol. 156 (G. Braziller, 1958).
2. Fee, J., 'Maupertuis, and the Principle of Least Action', *Scientific Monthly*, 52, 496–503 (1941).
3. Randall, L., and Reece, M., 'Dark Matter as a Trigger for Periodic Comet Impacts', *Physical Review Letters*, 112, 161301 (2014).
4. Carroll, S., 'Painting Pictures of Astronomical Objects', *Discover*; https://www.discovermagazine.com/the-sciences/painting-pictures-of-astronomical-objects#.WcJ-s8ZJnIU.
5. Rubin, V. C., and Ford Jr, W. K., 'Rotation of the Andromeda Nebula From a Spectroscopic Survey of Emission Regions', *Astrophysical Journal*, 159, 379 (1970).
6. Oaknin, D. H., and Zhitnitsky, A., 'Baryon Asymmetry, Dark Matter, and Quantum Chromodynamics', *Physical Review D*, 71, 023519 (2005).

7. Barrow, J. D., and Tipler, F. J., *The Anthropic Cosmological Principle* (Clarendon Press, 1986).
8. Smolin, L., *The Life of the Cosmos* (Oxford University Press, 1999).
9. Smolin, L., *Time Reborn: From the Crisis in Physics to the Future of the Universe* (Houghton Mifflin Harcourt, 2013).
10. Lloyd, S., *Programming the Universe: A Quantum Computer Scientist Takes on the Cosmos* (Knopf, 2006).
11. Vilenkin, A., 'Creation of Universes from Nothing', *Physics Letters B*, 117, 25–8 (1982).
12. Dennett, D. C., 'Darwin's Dangerous Idea', *Sciences*, 35, 34–40 (1995).

## Epilogue

1. Spade, P. V., *The Cambridge Companion to Ockham* (Cambridge University Press, 1999). Riezler, S., *Vatikanische Akten Zur Deutschen Geschichte in Der Zeit Kaiser Ludwigs Des Bayern* (Wentworth Press, 2018).
2. Hoffmann, R., Minkin, V. I., and Carpenter, B. K., 'Ockham's Razor and Chemistry', *Bulletin de la Société chimique de France*, 2, 117–30 (1996).
3. Wheeler, J. A., 'How Come the Quantum?', *Annals of the New York Academy of Sciences*, 480, 304–16 (1986).

# Index

Page numbers in *italics* denote figures

**Johnjoe McFadden** is professor of molecular genetics at the University of Surrey, where he studies the genetics of microbes that cause infectious diseases, such as tuberculosis. The author of *Quantum Evolution* and the coauthor of *Life on the Edge*, he lives in London with his wife and son.